D1384751

EYES TO THE SKY

EYES TO THE SKY

PRIVACY AND COMMERCE IN THE AGE OF THE DRONE

EDITED BY MATTHEW FEENEY

Print ISBN: 978-1-952223-08-2
eBook ISBN: 978-1-952223-09-9

Cover design: Molly von Borstel, Faceout Studio
Imagery: Shutterstock

Library of Congress Cataloging-in-Publication Data

Feeney, Matthew, editor.
Eyes to the sky : privacy and commerce in the age of the drone / edited by Matthew Feeney.
pages cm
Washington, DC : Cato Institute, 2021.
Includes bibliographical references and index.
ISBN 9781952223082 (hardcover) | ISBN 9781952223099 (ebook)
1. LCSH: Drone aircraft—Industrial applications. 2. Micro air vehicles—Government policy—United States. 3. Aerial surveillance—Government policy—United States. 4. Privacy, Right of—United States. 5. Air traffic rules—United States.
TL685.35 .E94 2021
629.133/39—dc23 2021017826

Printed in Canada.

CATO INSTITUTE
1000 Massachusetts Avenue NW
Washington, DC 20001

www.cato.org

CONTENTS

ABBREVIATIONS AND ACRONYMS vii

INTRODUCTION 1
Matthew Feeney

CHAPTER ONE 5
"Crawl, Walk, Fly": A History of UAS Regulation in the United States
Sara Baxenberg

CHAPTER TWO 53
Who Wants a Drone Anyway? The Law Develops to Accommodate the Promise of Commercial Drones
Gregory S. Walden

CHAPTER THREE 75
Reframing Drone Policy to Embrace Innovation in America
James Czerniawski

CHAPTER FOUR 111
Who Should Govern the Skies?
Brent Skorup

CHAPTER FIVE 131

Who Owns the Skies? Ad Coelum, *Property Rights, and State Sovereignty*

Laura K. Donohue

CHAPTER SIX 165

Legislative Rules for Use of Drones by Law Enforcement

Jake Laperruque

CHAPTER SEVEN 181

Drone Capabilities and Their Uses by the Federal Government

Jay Stanley

ACKNOWLEDGMENTS 209

NOTES 211

INDEX 273

ABOUT THE EDITOR 287

ABOUT THE CONTRIBUTORS 289

ABBREVIATIONS AND ACRONYMS

AAM	advanced air mobility
AGL	above ground level
AI	artificial intelligence
ALI	American Law Institute
ANPRM	Advance Notice of Proposed Rulemaking
ARC	Aviation Rulemaking Committee
ASTM	American Society for Testing and Materials
ATC	air traffic control
ATF	Bureau of Alcohol, Tobacco, Firearms and Explosives
ATM	air traffic management
BVLOS	beyond the visual line of sight
CAA	Civil Aeronautics Authority
CBP	U.S. Customs and Border Protection
COA	Certificate of Waiver or Authorization
COW	cell on wheels, cell on wings
DAA	detect and avoid
DHS	U.S. Department of Homeland Security
DJI	Da-Jiang Innovations
DOJ	U.S. Department of Justice
DOT	U.S. Department of Transportation
FAA	Federal Aviation Administration

FCC	Federal Communications Commission
FOIA	Freedom of Information Act
FRIA	FAA-recognized identification areas
GAO	U.S. Government Accountability Office
GSD	ground sample distance
ID	identification
IG	inspector general
IPP	Integration Pilot Program
LAANC	Low Altitude Authorization and Notification Capability
LEAP	Long Endurance Aircraft Platform
NAS	National Airspace System
NIMBY	not in my backyard
NPRM	Notice of Proposed Rulemaking
NTIA	National Telecommunications and Information Administration
OIRA	White House Office of Information and Regulatory Affairs
OOP	operations over people
PSS	Persistent Surveillance Systems
sUAS	small unmanned aerial system
UAS	unmanned aircraft system
UAV	unmanned aerial vehicle
ULC	Uniform Law Commission
UPP	UTM Pilot Program
UPS	United Parcel Service
USS	UAS service supplier
UTM	UAS traffic management
VADER	Vehicle and Dismount Exploitation Radar
WAMI	wide-area motion imagery
WASS	Wide Area Surveillance System

INTRODUCTION

MATTHEW FEENEY

For many years, the word "drone" prompted Americans to consider the multimillion-dollar weapon platforms used in the United States' ongoing and seemingly endless war on terror. More recently, drones—often referred to as "unmanned aerial vehicles"—have become a more common feature of civilian life, with a range of industries using them for inspections, photography, mapping, and much more. Law enforcement agencies have also expressed an interest in drones. The Department of Homeland Security has flown predator drones on the northern and southern borders for years, but as drones have become smaller and less expensive, they have become increasingly attractive to state and local law enforcement agencies. The commercial applications of drones have prompted lawmakers and regulators across the world to seek policies that allow for the innovative and potentially life-saving applications of drones while also protecting civil liberties from intrusive eyes in the sky. This volume, which brings together writers from academia, public policy, law, and civil liberties advocacy, explores such policies.

Although the Fourth Amendment protects against "unreasonable searches and seizures," the Supreme Court's treatment of aerial surveillance does not prohibit warrantless drone surveillance. Nonetheless, Justice William Brennan asked his colleagues to consider such devices in his dissent in a case from the 1980s considering warrantless surveillance from a helicopter. He wrote: "Imagine a helicopter capable of hovering just above an enclosed courtyard or patio without generating any noise, wind, or dust at all. . . . Suppose the police employed this miraculous tool to discover not only what crops people were growing in their greenhouses, but also what books they were reading and who their dinner guests were."[1] Today, flying machines similar to Brennan's "miraculous tool" are in the hands of police departments across the world.

The Supreme Court has yet to consider whether warrantless drone surveillance violates the Fourth Amendment, but that has not stopped lawmakers across the country from imposing warrant requirements for drone surveillance. Chapters Six and Seven explore the surveillance capabilities of drones and the steps lawmakers at the federal and state levels can take to mitigate the threat of drone surveillance. This issue is of especially urgent concern at a time when police departments across the country have demonstrated an eagerness to expand their aerial surveillance capabilities and drone surveillance technology is improving.

Even if lawmakers were able to adequately protect us from snooping drones, difficult questions would remain over how best to integrate drones into a regulated airspace designed for traditional aircraft. Drones have many benefits, but there are risks associated with drone collisions, drones falling out of the sky, and drones interacting with airport infrastructure.

In chapters on the commercial use and regulation of drones, this volume's contributors propose policies that aim to make American airspace more accommodating to drones while highlighting the range of uses for these vehicles. Chapter One focuses exclusively on the history of the Federal Aviation Administration's efforts to regulate drones.

Although drones are a relatively new technology, they raise legal questions that have been discussed for hundreds of years, such as "Who owns the airspace above property?" Chapter Five traces the long history of how common law has treated airspace and the way in which that law continues to influence how lawmakers and judges think about questions concerning aerial trespass and nuisance in the 21st century.

As you read the following pages, you will learn about predator drones, King Edward I's role in the development of aviation law, air taxis, atmospheric satellites, and how frightened chickens on a North Carolina farm led to a crucial Supreme Court decision. Perhaps more importantly, you will understand the wide range of risks and opportunities posed by the emergence of the drone and how, with the right policies in place, lawmakers and regulators can protect our civil liberties while ensuring that hobbyists, businesses, artists, and many others can take advantage of an exciting new technology.

CHAPTER ONE

"CRAWL. WALK. FLY": A HISTORY OF UAS REGULATION IN THE UNITED STATES

SARA BAXENBERG

The ongoing development of regulations for unmanned aircraft systems (UAS or "drones") in the United States demonstrates the challenges of integrating a new technology into a heavily regulated field. These challenges have been particularly significant given the complexity of the U.S. National Airspace System (NAS), the Federal Aviation Administration's (FAA) mandate to ensure aviation safety, and society's deep-seated aversion to aviation-related accidents. Although regulatory progress has ultimately been much slower than either the FAA or the UAS industry anticipated, the agency has made substantial strides in the decade since receiving its first mandate to integrate unmanned aerial vehicles (UAVs) into the airspace. The lessons learned in this time may—and hopefully will—lead to faster progress over the next decade.

THE REGULATORY STARTING POINT: GROUNDED BY DEFAULT

In 2012, when the FAA received its legislative mandate to integrate UAS into the national airspace, the agency was faced with a monumental task. While operating under a robust and largely incompatible existing regulatory environment, the FAA needed to find a way to enable scalable UAS operations in a complex airspace, all while maintaining the United States' unparalleled aviation safety record. Because UAS did not fit easily into preexisting regulatory categories, the legal landscape into which the UAS industry was born, in the United States at least, was one of grounding by default: UAS operations were generally prohibited until the FAA could find a way to permit them. To change that and get UAS off the ground while protecting existing airspace users, the FAA has referred to its regulation of UAS operations as requiring a "crawl, walk, run" approach.[1] The metaphor is accurate, but the process has involved far more crawling than either the industry or the FAA expected.

The challenge of integrating UAS into the national airspace is best understood by examining the regulatory environment that existed when consumer drones started to proliferate. The FAA was created in 1958 to coordinate the operations of military and civil aircraft after a series of high-profile accidents. More than 40 years later, in the years leading up to the 2012 congressional mandate, the United States boasted both the most complex, active airspace in the world and an exceptional safety record. To reach this point, the FAA had instituted expansive and complex regulatory frameworks, governing areas such as operational authorization, the design and equipage of aircraft, pilot qualifications, flight routes, and coordination with air traffic control.

Thus the regulatory gauntlet facing the civilian UAS industry long predated the industry's very existence. Unpiloted remote-control aircraft, in a variety of shapes and sizes, have been built and flown by hobbyists for decades, generally without interfering with other users of the NAS. These aircraft were considered recreational and categorized as "model aircraft." Because of the long history of self-regulation by this small subset of airspace users and the limited scope of their operations, the FAA allowed them to operate largely unregulated, their operations governed by a sparse advisory circular that did not apply to commercial operators.[2] However, the development of new technologies—such as powerful batteries, inexpensive gyroscopes, and computerized flight controls—presented new possibilities and uses for unmanned aircraft. These technologies also opened flying up to a much larger slice of the public, who could purchase extremely capable unmanned aircraft off the shelf and fly them with little or no practice. Commercial entities soon began to identify applications for small UAS that could obviate the need for activities on foot or in manned aircraft. Deploying drones for dangerous and expensive tasks could expand companies' existing capabilities and generally transform numerous aspects of how we live and work. It rapidly became clear that a regulatory framework was necessary to enable these large-scale commercial UAS operations and that the existing regulatory regimes were not a good fit.

By their nature, small UAVs could not meet regulatory requirements designed to ensure the safety of manned aircraft. For instance, they could not be equipped with fire extinguishers or safety manuals on board: they could not be "boarded" at all, and the small size and battery-powered nature of consumer UAVs made including even the slightest unnecessary equipage infeasible.[3] Similarly, while FAA regulations defined "navigable airspace" as starting at altitudes of 500 or 1,000 feet and generally required aircraft to operate above that threshold other than on takeoff and landing, unmanned aircraft were designed to operate at—and indeed, would maximize safety and efficiency at—significantly lower altitudes.[4] And the complex regulatory processes surrounding aircraft certification and pilot training,[5] which imposed justifiable burdens on large aircraft with significant capacity for cargo or people, were economically infeasible for a $1,000 aircraft with a 30-minute flight range and a mere 10-pound payload. Moreover, these processes were far more onerous than was necessary to ensure safety given the significantly lower risk posed by aircraft weighing only a handful of pounds and operating close to the ground.

The grounding by default that resulted from the inability to fit UAS into existing aviation regulation frameworks presented a significant challenge to UAS integration. It also stands in stark contrast to other areas where new technology has been able to flourish. Innovations such as smartphones, automated vehicles, and even manned aviation itself were born into regulatory environments that enabled rapid development and widespread deployment followed by some degree of regulatory backlash. In contrast, drones have largely struggled to get off the ground at all. This regulatory positioning has undoubtedly contributed to a lag in public acceptance and continues to threaten the success of the industry.

Nonetheless, given the FAA's development of policies and programs to make small UAS flights possible, UAS integration has begun in earnest. Furthermore, the industry stands poised to enter a regulatory era that will enable widespread proliferation of UAS use and applications over the next few years. This chapter walks through the significant developments that have enabled the UAS industry to reach this point in its development, including

- Congress's 2012 mandate to the FAA to integrate UAS into the airspace in the FAA Modernization and Reform Act,
- the FAA's case-by-case authorization process for commercial operations under Section 333 of the act,
- the FAA's adoption of its Part 107 regulations to broadly enable commercial UAS operations subject to a number of operating limitations,
- the need for technology to enable remote identification of UAVs in flight before the FAA could enable widespread expanded UAS operations, and
- the major steps the FAA has taken in between and since to keep the industry moving forward and build a regulatory structure that can support a truly integrated airspace in the future.

During the course of this nearly decadelong journey, the federal government has used a wide variety of tools to facilitate UAS integration, including legislation, case-by-case authorization and regulatory waivers and exemptions, federal advisory and rulemaking committees, pilot programs and public-private partnerships, emergency regulatory procedures, and traditional notice-and-comment rulemaking. This period has provided important lessons about legislative carveouts, interagency coordination, and the role of industry in informing the regulatory process. These lessons will undoubtedly shape—and, if all goes well, expedite—future UAS regulation.

THE PREREGULATORY PERIOD (2012–2016): LEGISLATION AND SECTION 333 EXEMPTIONS

THE ENABLING LEGISLATION

In the years approaching 2012, with the domestic UAS industry in its nascent stages, drones began to emerge at price points consumers could afford and with capabilities that would open up a new universe of commercial and recreational activities. And yet the FAA's pervasive regulation of the national airspace had, to date, been built expressly for manned aircraft.

The ability of drones to use the airspace even in a limited way would depend on significant relief from existing regulatory frameworks, and widespread use likely would require new frameworks. Although the FAA has general authority to waive its regulatory requirements,[6] Congress recognized that efficient and effective integration of UAS into the national airspace system required a more comprehensive effort and regulatory change. The first major legislative action toward UAS integration came at the end of 2011 with the passage of the National Defense Authorization Act for Fiscal Year 2012 (2012 NDAA). The 2012 NDAA directed the FAA to "establish a program to integrate [UAS] into the national airspace system at six test ranges."[7] To implement this program, the FAA was required to (a) designate nonexclusionary airspace for UAS operations, (b) develop certification standards and air traffic requirements for test range operations, (c) "coordinate with and leverage the resources of the Department of Defense and the National Aeronautics and Space Administration," (d) address both public and privately operated UAS, (e) coordinate the program with the FAA's NextGen airspace modernization project, and (f) "provide for verification of the safety of [UAS] and related navigation procedures before integration into the national airspace system."[8] Consistent with this mandate, the FAA established six test sites in 2013 by approving applications from the University of Alaska, the State of Nevada, New York's Griffiss International Airport, the North Dakota Department of Commerce, Texas A&M University at Corpus Christi, and the Virginia Polytechnic Institute and State University (Virginia Tech).[9]

The landmark legislation in the area of UAS integration—the 2012 FAA Modernization and Reform Act ("FMRA")—followed closely on the heels of the 2012 NDAA. The FMRA was Congress's first full-throated attempt to provide legislative direction to enable widespread UAS integration (some provisions would prove more successful than others). Section 332 of the act, Integration of Civil Unmanned Aircraft Systems into the National Airspace System, imposed a number of requirements on the FAA in furtherance of that objective. These included (a) the creation of a comprehensive plan to "safely accelerate" UAS integration, developed "in consultation with representatives of the aviation industry, Federal agencies that employ unmanned aircraft systems technology in the national airspace system, and the unmanned aircraft systems industry";[10] (b) the creation of a five-year roadmap to accompany the plan;[11] (c) the promulgation of a final rule within 18 months to permit civil operation of small UAS in the NAS;[12] and (d) the promulgation of a final rule within 34 months to implement certain recommendations from the comprehensive plan, such as standards for operation and certification of civil UAS, ensuring that UAS include sense-and-avoid capabilities, and registration and licensing standards for pilots.[13] Recognizing that some UAS operations likely could be safely allowed before the completion of a rulemaking process, Congress also empowered the secretary of transportation, under Section 333 of the FMRA, to make case-by-case determinations as to whether specific UAS operations could safely be conducted in the national airspace, and to set the criteria for such operations.[14]

Section 333 was essential to UAS integration because it authorized the FAA to circumvent federal law requiring aircraft operators to hold an airworthiness certificate.[15] An airworthiness certificate indicates that an aircraft conforms to the relevant type certificate (a separate FAA certification approves the design and manufacture of an aircraft) and is in safe condition for flight.[16] To this point, federal law expressly prohibited "operat[ing] a civil aircraft in air commerce without an airworthiness certificate."[17] Since 2005, the FAA had been working on developing regulations for UAS but had ultimately determined that exempting UAS from the statutory requirement to obtain an airworthiness certificate would be necessary to enable scalable civil operations. However, the agency lacked the necessary statutory authority to do so.[18] Section 333 provided that mechanism. The FAA ultimately would rely on the authority granted by Section 333 to adopt regulations that enable commercial UAS operations without an airworthiness certificate.

In addition to addressing the use of UAS as civil aircraft—defined by preexisting statute as all aircraft that are not "public," meaning government owned or operated—the FMRA also directed the secretary of transportation to issue guidance to expedite authorization of public UAS, facilitate public agencies' ability to use UAS test sites, clarify public entities' responsibilities while operating UAS without civil airworthiness certificates for the aircraft, and establish a "collaborative process" that would "allow for an incremental expansion of access to the national airspace system as technology matures and the necessary safety analysis and data become available. . . ."[19] The secretary of transportation was also required to enter into agreements with "appropriate government agencies" to simplify the process for granting Certificates of Waiver or Authorization (COAs) for public UAS use, and the FAA was required to "develop and implement operational and certification requirements" for public UAS operations.[20]

Finally, the FMRA included a provision that divested the FAA of regulatory authority over a subset of civil UAS operators: those flying "model aircraft" or, in other words, hobbyists. Specifically, Section 336 of the act provided that the FAA "may not promulgate any rule or regulation regarding a model aircraft, or an aircraft being developed as a model aircraft" if "the aircraft is flown strictly for hobby or recreational use."[21] The statute included other criteria that a model aircraft must meet to be outside the purview of the FAA's regulatory authority, including that the aircraft be "operated in accordance with a community-based set of safety guidelines and within the programming of a nationwide community-based organization," weigh no more than 55 pounds "unless otherwise certified through a . . . program administered by a community-based organization," be operated so as to give way to manned aircraft, and operate within five miles of an airport only after the operator has provided notice to the airport operator and air traffic control (ATC) tower, if applicable.[22] The definition of "model aircraft" provided further clarity on the scope of the exemption, applying to aircraft "capable of sustained flight in the atmosphere" and "flown within visual line of sight" of the operator.[23] The only limitation on this carveout of FAA authority centered around reckless operation: Section 336 did not limit the authority of the FAA to "pursue enforcement action against persons operating model aircraft who endanger the safety of the national airspace system."[24] Although this section was adopted with an eye toward the model aircraft hobbyists belonging to long-standing organizations such as the Academy of Model Aeronautics, which had roughly 150,000 members around the time the FMRA was passed,[25] Section 336 became enormously consequential for the UAS industry at large.

SECTION 333 EXEMPTIONS

With the test sites up and running and allowing experimentation with various UAS applications and operations, the FAA moved on to the next step in its plan to integrate UAS into the national airspace: Section 333 exemptions. From September 2014 until mid-2016, when the FAA adopted the broadly enabling regulations required by Section 332 of the FMRA, the Section 333 process was the sole means to obtain authorization to conduct commercial UAS operations.

As set forth previously, Section 333 directed the FAA to make case-by-case determinations as to whether specific UAS operations could safely be conducted in the national airspace, notwithstanding the lack of enabling regulations. Entities seeking to operate UAS pursuant to this process were required to submit petitions for exemption with detailed information about the nature of the operations and aircraft and the FAA regulations from which relief was needed to conduct those operations. Applicants generally provided significant documentation to support their claims that the operations could be conducted safely, including manufacturer operations manuals and aircraft specifications documents, pilot training programs, and preflight checklists.

If the FAA granted the operator's petition, the resulting grant generally imposed about three dozen operating conditions, including speed and altitude limitations (typically 400 feet above ground level) and requirements that operations use a visual observer and that the UAS stay within visual line of sight of both the pilot and the observer.[26] Some of the Section 333 grant conditions were particularly onerous. For example, operations had to be conducted at least 500 feet from nonparticipating persons, at least five miles from an airport, and at least three miles from any "city or densely populated area," and the UAS pilot had to hold a pilot certificate. The latter condition required being certificated in piloting some type of manned aircraft, as no UAS-specific pilot's license existed at the time. In addition, Section 333 grants were limited to the operations and aircraft described in the application, thus necessitating a new petition and an amended grant to operate new aircraft or for a new purpose. Finally, the grant was only one of two authorizations that a Section 333 petitioner needed to fly in the national airspace: each Section 333 grant required the applicant to also obtain a COA to conduct operations. The COA, in turn, had its own limitations. Each COA was limited to specified coordinates and required the operator to issue a Notice to Airmen between 24 and 72 hours prior to operation, thus significantly constraining both the timing of operations and the operating area.

Accordingly, although the Section 333 process authorized thousands of commercial operations in the two-year period before the FAA adopted broader enabling regulations,[27] those authorizations were exclusively case by case; the application process was burdensome; agency processing times were long (on the order of months); and the authorized operations were constrained with respect to location, aircraft, pilot qualifications, and operating restrictions. However, there were efficiency gains as the process developed: once the FAA began granting exemptions and the petitions (though not the supplementary documentation) were made public, subsequent operators were able to leverage the work done by the early movers with respect to the form and substance of the applications. In addition, the FAA took several steps during the Section 333 era to streamline the process and improve operational flexibility. Those steps included (a) introducing a "blanket COA," which would be summarily granted to all Section 333 grant recipients and enable operations anywhere in the country other than in the vicinity of airports up to an altitude of 200 feet, thereby eliminating the need for an operator to obtain its own COA unless the requested operations exceeded that altitude;[28] (b) moving to a summary grant process whereby the FAA would grant applications without publishing a notice in the *Federal Register* or undertaking a detailed analysis if the agency had already granted a similar request, which decreased typical Section 333 processing time from months to several weeks;[29] and (c) adopting a list of several hundred "approved UAS" that Section 333 grantees could use, thus in many cases obviating the need to file a new petition to add aircraft to an existing grant.[30] Nonetheless, case-by-case authorization was clearly not a sustainable long-term model for UAS integration for either the industry or the FAA. The UAS space needed affirmative, enabling regulations.

PART 107: THE FAA'S FIRST FORAY INTO ENABLING REGULATIONS FOR DRONES

Congress anticipated that, given the complexity of the National Airspace System and FAA regulatory frameworks, rules that broadly enabled UAS operations would be necessary. As discussed previously, this foresight was reflected in Section 332 of the FMRA, which required the FAA to adopt "a final rule on small unmanned aircraft systems that will allow for civil operation of such systems in the national airspace."[31] The agency satisfied that statutory mandate in June 2016, more than four years after the passage of the FMRA, when it adopted its Part 107 regulations.

Taking shortly over one year from Notice of Proposed Rulemaking (NPRM)[32] to final rule (relatively speedy by agency rulemaking standards), Part 107 enabled routine—though limited—operations of small (below-55-pound) UAS without the need for prior FAA authorization. Many of the operating limitations adopted in the Part 107 regulations mirrored the constraints of typical Section 333 grants, including a 400-foot altitude limitation; speed limit; visibility requirements; and prohibitions on conducting operations at night, over people, and beyond the visual line of sight (BVLOS).[33] However, Part 107's operating conditions were (and remain) less stringent than their predecessor Section 333 requirements in a number of respects: operators under Part 107 are not required to file a Notice to Airmen prior to flight, need not maintain a specific distance from cities or densely populated areas, are not restricted to any particular aircraft, and do not need to obtain COAs. In addition, in adopting Part 107 the FAA tackled its pilot certification problem, creating a new certification (Remote Pilot with Small UAS Rating) more appropriately tailored to the knowledge, skills, and level of experience required to fly drones. As compared with pilot certification requirements for manned aircraft, the remote pilot certification process requires a less difficult test and does not have a flight hours component, although a background check by the Transportation Security Administration is still required.[34] This development alleviated a significant operational burden, as finding a certificated pilot or having a prospective UAS pilot become certificated as a pilot of manned aircraft often created an operational bottleneck for operators. As of July 2018, the FAA had issued more than 100,000 remote pilot certificates and reported a success rate for the knowledge test of 92 percent.[35]

PART 107'S PUNT ON AIR CARRIERS

In adopting the Part 107 rules, the FAA declined to wade into one major area of interest to the burgeoning drone industry: air carrier operations. Although many companies at the time had already publicly introduced plans for large-scale package delivery programs, the existing statutory and regulatory framework made the endeavor relatively more complicated in practice than other types of UAS operations. Pursuant to the Federal Aviation Act, entities that qualify as "air carriers"—defined as persons engaged in "air transportation," or "the transportation of passengers or property by aircraft as a common carrier for compensation"[36]—are required by statute to obtain three separate certifications: (a) an air carrier operating certificate premised on compliance with safety requirements for the specific operations;[37] (b) an airworthiness certificate designed to ensure that the aircraft to be used meets relevant design, construction, and performance requirements;[38] and (c) a certificate authorizing the air transportation premised on compliance with economic requirements, such as holding proper liability insurance and establishing reasonable rates.[39] Because the act requires the FAA to consider both "the duty of an air carrier to provide service with the highest possible degree of safety in the public interest" and the "differences between air transportation and other air commerce" when prescribing regulations governing the operations of air carriers, the safety regulations applicable to air carriers are some of the most detailed, lengthy, and robust rules promulgated by the agency.[40]

Rather than attempt to navigate the complexities of the statutory and regulatory framework applicable to air carriers so as to apply it to UAS, the agency punted on the issue. It drafted its Part 107 rules specifically to avoid the air carrier framework—and to significantly restrict package delivery operations in the process. Specifically, Section 107.1, which defines the scope of the rules, states that Part 107 does not apply to "air carrier operations."[41] In the narrative adopting the rules, the agency explained that this would permit UAS operations involving the transportation of people or property that are wholly intrastate, as the statutory definition of "air transportation" enumerates three types of such transportation: interstate, foreign, and transport of mail. Accordingly, UAS operators could engage in the transport of property if such transport "occur[red] wholly within the bounds of a state," and thus could utilize Part 107 while avoiding air carrier designation.[42] The agency's intent in structuring the rule this way was to "provide immediate flexibility for remote pilots to engage in the limited carriage of property by small UAS," while giving the FAA the opportunity to "evaluate the integration of more expansive UAS air carrier operations into the NAS and . . . propose further economic and safety regulations if warranted."[43]

Notably, although requiring UAS operations to remain intrastate appears to be a straightforward way to avoid the statutory air carrier classification that applies to entities providing "interstate air transportation," avoiding air carrier status may be more difficult than it appears, as the case law demonstrates that "the terms 'air carrier' and 'air transportation' apply more broadly than one might guess."[44] For instance, the U.S. Ninth Circuit Court of Appeals found that the wholly intrastate trucking legs of FedEx's delivery network constitute air transportation because they are an "essential part of [FedEx's] all-cargo air service."[45] Accordingly, the court held that state economic regulation of the trucking business was preempted by the Airline Deregulation Act of 1978, which prohibits states and localities from "enact[ing] or enforc[ing] any law, rule, regulation, standard, or other provision . . . relating to rates, routes or services of any air carrier."[46] The FAA did not contend with these legal nuances in the Part 107 rule, and as of early 2021 the issue does not appear to have come up in the courts with respect to intrastate UAS flights that are part of an interstate delivery system.

THE PART 107 WAIVER PROCESS

The FAA's Part 107 rulemaking also evinces the agency's acknowledgment of the operating limitations the rules impose and its desire to facilitate safe operations that deviate from those limitations. In adopting the rules, the agency observed that both "(1) the rulemaking process for higher-risk UAS operations may lag behind new and emerging technologies; and (2) certain individual operating environments may provide unique mitigations for some of the safety concerns underlying this rule."[47] Accordingly, the FAA adopted a waiver process to allow for case-by-case relief from most of Part 107's operating limitations. Specifically, the following rules under Part 107 can be waived:

- Section 107.25 – Operation from a moving vehicle or aircraft (except where waiver would permit carriage of property of another by aircraft for compensation or hire, thereby making the UAS operator an air carrier as discussed earlier)
- Section 107.29 – Daylight operation
- Section 107.31 – Visual line of sight aircraft operation (except where waiver would permit carriage of property of another by aircraft for compensation or hire, thereby making the UAS operator an air carrier as discussed earlier)
- Section 107.33 – Visual observer
- Section 107.35 – Operation of multiple small unmanned aircraft systems
- Section 107.37(a) – Yielding the right of way
- Section 107.39 – Operation over people
- Section 107.41 – Operation in certain airspace
- Section 107.51 – Operating limitations for small unmanned aircraft[48]

The FAA's intent was that the waiver process would allow relief from Part 107 on a timeline of 90 days rather than the longer processing time frames typically required for Section 333 exemptions, and that the adoption of an online portal for submitting waiver requests and receiving FAA responses would make the process smooth and efficient. In furtherance of that effort, the FAA issued Waiver Safety Explanation Guidelines to aid applicants in determining the operational risks and requisite safety mitigations associated with waiver from a given rule[49] and held a series of webinars on how to create successful waiver applications for specific types of operations.[50] Waivers from Section 107.41 governing operations in controlled airspace were handled through a different application form than other Part 107 waivers and, as discussed in more detail later, were ultimately replaced by a real-time airspace authorization system (the Low Altitude Authorization and Notification Capability, or LAANC) operated by third-party service suppliers. For operations that deviate from Part 107 rules that are not waivable—for instance, those operations involving aircraft over 55 pounds—the Section 333 process would continue to be available.

The FAA's Part 107 waiver program has had mixed results. As of December 2019, the FAA had granted approximately 3,500 waivers from Part 107 (other than controlled airspace authorizations).[51] However, the vast majority of these authorizations (nearly 3,300) provided relief from Section 107.29 to enable nighttime operations. These waivers were relatively straightforward, requiring that the operations use a visual observer, that the UAS be equipped with anti-collision lighting and the operating area be sufficiently illuminated, and that the pilot be trained on the nuances of operating UAS in the dark.[52] In contrast, by this time the agency had issued fewer than 100 waivers to allow flights over persons, and fewer than 50 each to enable BVLOS operations and simultaneous operation of multiple UAS. A 2018 report by the U.S. Department of Transportation Office of Inspector General evaluating the FAA's Part 107 waiver process found that although the agency's Flight Standards Office "consistently met its established goal to review 80 percent of waiver applications within 90 days," it had "disapproved nearly three-quarters of waiver applications (73 percent), primarily due to incomplete information or an insufficient safety case."[53] Thus, waiver requests for the more complex operations (those requiring waiver from rules other than just the prohibition on nighttime operations) on average required longer wait times for approval—with only 4 percent of such waivers being approved within the FAA's aspirational 90-day processing time frame.[54] Airspace waivers fared even worse, with 66 percent of roughly 9,000 waiver requests still pending review as of release of the report.[55] Ultimately, the report concluded that the FAA's experience with the waiver process demonstrates that the agency "is still challenged to collect UAS data needed to assess potential safety hazards, educate non-traditional aviators on its safety culture, and effectively implement a risk-based oversight system."[56] This challenge, the inspector general found, "hinders the Agency's ability to accurately measure and mitigate the safety risks UAS could pose and transition effectively from waiver-based operations to full integration into the NAS."[57]

REGULATORY DELAYS FOR EXPANDED OPERATIONS

Once the Part 107 rules went into effect in August 2016, commercial UAS operators had unprecedented access to airspace (well, some airspace) without prior FAA authorization. However, Part 107's significant operating limitations—which prohibit operations over people, at night, and BVLOS, among other restrictions—continued to pose obstacles for operators seeking to undertake large-scale operations for purposes such as infrastructure inspection, insurance appraisal, package delivery, and numerous others. Although the streamlined waiver process was somewhat easier to navigate and less burdensome than the Section 333 exemption process, it was clear that enabling regulations like those adopted in Part 107 would be necessary to allow these expanded operations to be scaled and meaningfully deployed.

The FAA was well aware that further enabling regulations were needed. Part 107 was just "one step of a broader process to fully integrate UAS into the NAS."[58] In developing Part 107, the FAA made the policy decision that it was best to "proceed incrementally and issue a final rule that immediately integrates the lowest-risk small UAS operations into the NAS" while the agency "continue[d] working on integrating UAS operations that pose greater amounts of risk."[59] To organize and illustrate this incremental approach, the FAA developed a chart that identified the regulatory actions it would need to take to achieve full UAS integration, which it often displayed at industry conferences. In September 2016, the chart appeared as in Figure 1, though it was updated relatively frequently as the FAA's policy vision evolved.

FIGURE 1
FAA UAS integration strategy

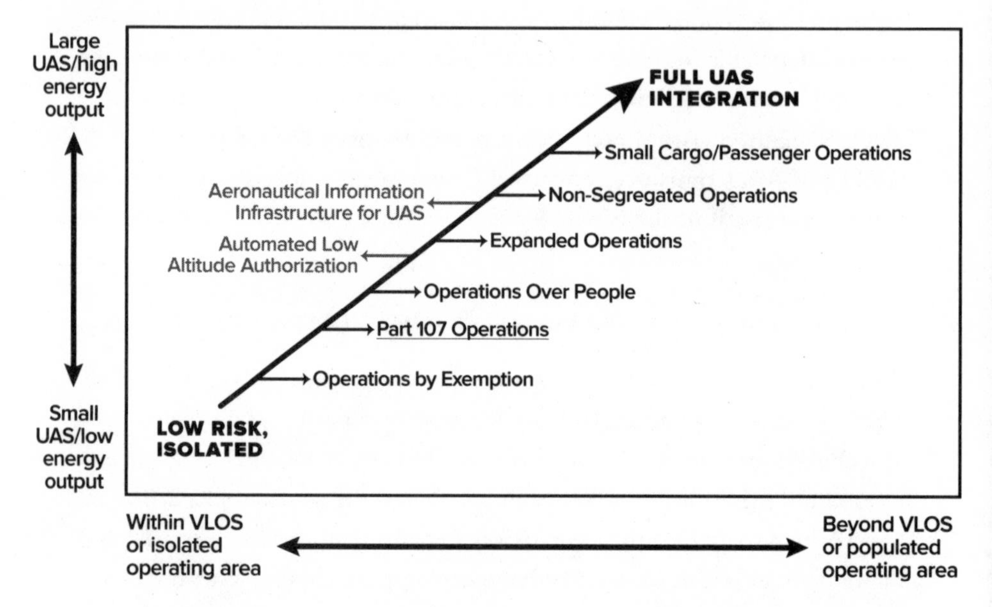

Source: Earl Lawrence, "Drone Advisory Committee Public Meeting: Overview of the UAS Landscape," FAA Drone Advisory Committee Public Meeting, Washington, September 16, 2016, at 13, https://www.faa.gov/uas/programs_partnerships/drone_advisory_committee/rtca_dac/media/dac_sept_2016_ppt_presentation.pdf.
Notes: Figure reproduced from original. FAA = Federal Aviation Administration; UAS = unmanned aircraft system; VLOS = visual line of site.

Shortly after this slide was presented at a meeting of a UAS-focused FAA advisory committee, and just a few short months after Part 107 became effective, the FAA reportedly was preparing to take the next big step on its incremental approach chart: enabling flights over people.

FLIGHTS OVER PEOPLE IN 2017: THE PHANTOM NPRM

For the FAA, flights over people served as a reasonable starting point to expand commercial UAS operations beyond Part 107 for two reasons. First, the inability to operate over people placed significant restrictions on the burgeoning UAS industry. Second, the agency already had a head start on tackling this problem. In the NPRM proposing the rules that were adopted as Part 107, the FAA proposed creating a "micro UAS" category that would enjoy relaxed regulatory restrictions given the relatively lower risks posed by aircraft of that size.[60] Under the FAA's proposal, unmanned aircraft weighing less than 4.4 pounds and constructed of frangible materials could be operated in Class G (noncontrolled) airspace over people, and the operator could obtain a less stringent pilot certificate (micro UAS rating rather than small UAS rating) that would not require a knowledge test.[61]

The FAA based these proposals on regulatory approaches to micro UAS in other countries as well as recommendations issued in 2009 by the agency's first UAS-related Aviation Rulemaking Committee (ARC)—the Small UAS ARC. The FAA uses ARCs to solicit industry feedback on aviation-related issues in an organized way as a precursor to rulemaking. The Small UAS ARC first issued recommendations to the FAA on integrating UAS into the NAS in April 2009, well before the passage of the FMRA and the FAA's release of the NPRM proposing the rules that ultimately would become Part 107.

Despite teeing up the issue of micro UAS that could fly over people in the Part 107 NPRM, the agency ultimately concluded that its proposal was simultaneously too restrictive and potentially not restrictive enough in light of available data. Indeed, in the final rule the FAA agreed with stakeholders both that "the micro UAS limitations proposed in the NPRM, such as the requirement to remain more than five miles away from an airport and the prohibition on autonomous operations would, if finalized in this rule, significantly impair micro UAS operations" and that "even though micro UAS are smaller than other small UAS, they can still pose a safety risk," a concern that "is particularly troubling given the limited safety data currently available with regard to micro UAS operations and the fact that almost all other countries that currently regulate UAS generally do not allow small unmanned aircraft to fly over people or congested areas."[62] Accordingly, the FAA found that a "different framework" for micro UAS would be appropriate and would be undertaken in a separate proceeding.[63]

A few months before issuing the Part 107 final rule, the FAA chartered a new ARC to look specifically at micro UAS and UAS operations over people (the Micro UAS ARC). The NPRM that the agency had drafted in late 2016 was rumored to be based on the recommendations of the Micro UAS ARC, which had provided a report to the FAA in March 2016. The ARC proposed establishing four different categories of UAS based on the level of risk presented, with increasingly stringent performance standards and operating limitations.[64] Category 1 UAS would include all UAVs weighing 250 grams (0.55 pounds) or less and would be subject to no additional standards or restrictions.[65] At the other end of the spectrum, Category 4 would describe UAVs weighing more than 250 grams, whose operations require sustained flights over crowds, and that, upon impact with a person, pose less than a 30 percent chance of serious (Level 3) injury on the Abbreviated Injury Scale, an injury severity scoring system developed by the Association for the Advancement of Automotive Medicine. The Micro UAS ARC recommended that, to operate, Category 4 UAS be required to develop, in accordance with industry consensus standards, a risk mitigation plan specific to the operations and to comply with minimum stand-off distances from persons.[66]

By November 2016, the FAA had drafted and reportedly was planning to release an NPRM to expand Part 107 to enable flights over people. The FAA went as far as to send the NPRM to the Office of Information and Regulatory Affairs (OIRA) within the White House Office of Management and Budget for review, a necessary step in executive branch agencies' rulemaking process that usually occurs late in the process, once the agency has developed a complete draft of the item.[67] The NPRM was expected to be released to coincide with the Consumer Electronics Show in January 2017 and thus give the agency the opportunity to release important news at the heavily attended trade show. However, the show came and went, and the NPRM was not released.

THE REMOTE IDENTIFICATION BOTTLENECK

When the flights over people NPRM was not published as anticipated in January 2017, FAA staff familiar with the issue cited the need to address "stakeholder concerns"—specifically, "safety and security" concerns—before an NPRM on flights over people could move forward. In his remarks at the Consumer Electronics Show, then-FAA Administrator Michael Huerta delivered a consistent message, asserting that the agency had "been working diligently on a proposed rule. . . ." But "[a]llowing unmanned aircraft to fly over people raises safety questions because of the risk of injury to those underneath in the event of a failure." Huerta explained that UAS flights over people also "raise[d] security issues," providing by way of example "the challenge of a local police officer at a parade trying to determine which drones are properly there to photograph the festivities—and which may be operated by individuals with more sinister purposes." Addressing specifically the delay of the NPRM, Huerta explained, "The process of working with [the FAA's] interagency partners to reconcile these challenges is taking time." He also cited "issues" raised in "meetings conducted with industry stakeholders" but did not provide further detail.[68]

As time went on, it became clear that the safety and security concerns centered around the rules or standards related to the remote identification (ID) of UAVs in flight. The thinking within the federal government and in the industry was that if governmental entities such as law enforcement and security agencies had a means by which to determine the operator or owner of a drone—or to obtain other identifying information about the aircraft, such as registration number—while the aircraft was in flight, those entities would be better able to mitigate threats posed by UAS operating over people.

Remote ID had already emerged as a concept in the UAS space in 2016 in the FAA Extension, Safety, and Security Act (2016 Extension Act). Section 2202 of the act required the FAA—in consultation with the federal Department of Transportation (DOT), RTCA (a nonprofit entity that works with the FAA to convene stakeholders), and the National Institute of Standards and Technology—to convene industry stakeholders and develop "consensus standards for remotely identifying operators and owners of [UAS]."[69] In June 2017, the FAA chartered its third UAS ARC to develop recommendations on "remote identification and tracking" of UAS (Remote ID and Tracking ARC).[70] Specifically, the FAA asked the ARC to (a) recommend available and emerging remote ID and tracking technology for UAS; (b) identify requirements for meeting security and public safety needs of law enforcement, homeland defense, and national security communities with respect to remote ID and tracking; and (c) evaluate the feasibility and affordability of available technical remote ID and tracking solutions in light of the needs of both law enforcement and air traffic control.[71]

The Remote ID and Tracking ARC provided its recommendations to the FAA in September 2017. The committee identified two major categories of remote ID technology that bore further exploration by the FAA: (a) direct broadcast, pursuant to which the UAS would transmit data in one direction at regular intervals to no particular recipient, allowing anyone within range to receive the information; and (b) network based, pursuant to which the UAS would transmit the identifying information over a communications network, and those with authorized access to the network could obtain the information.[72] The ARC recommended that the information to be transmitted should include a unique identifier for the aircraft as well as the position of the aircraft and control station. Furthermore, it proposed that information identifying the UAS owner and remote pilot should not be transmitted but, instead, should be made available to authorized users through a system that could cross-reference the aircraft identifier with a separate, secure FAA database.[73]

A major issue of contention among the 74 stakeholders who made up the Remote ID and Tracking ARC was how broadly the rules should apply. Although the majority of stakeholders agreed that remote ID and tracking equipage requirements should not apply either to UAS operated within visual line of sight or to those that are incapable of operating BVLOS, some stakeholders favored a weight-based approach, pursuant to which all UAVs above a certain weight would be subject to remote ID and tracking requirements. Additionally, there was significant disagreement over whether the requirements should exempt "model aircraft." As set forth above, Section 336 of the 2012 FMRA established model aircraft as a category of UAS operated for hobby or recreational purposes that meet a series of criteria related to operational parameters and aircraft capabilities. The ARC report recommended that the FAA consider exempting from any remote ID and tracking requirements all model aircraft that comply with Section 336 and the FAA's corresponding rules (Part 101); the relevant underlying working group report indicated that available remote ID technologies "may be burdensome" for recreational operators.[74] Dissenters argued that exempting model aircraft from the requirements would be "a loophole that swallows the rule," enabling "a huge segment of the UAS community to avoid participating in the

UAS ID and Tracking system and complying with the corresponding . . . regulations."[75]

DRONE REGISTRATION AND THE CHALLENGE OF THE SECTION 336 HOBBYIST EXEMPTION

To accommodate existing hobbyist uses, Section 336 of the 2012 FMRA divested the FAA of regulatory authority over model aircraft, prohibiting the FAA from "promulgat[ing] *any* rule or regulation regarding a model aircraft."[76] The dispute at the Remote ID and Tracking ARC centered not on the FAA's authority to apply remote ID and tracking rules to model aircraft, but on the propriety of doing so. But the scope of Section 336—and the threat it posed to broad-based remote ID rules and other potential safety regulations—was a separate problem that would need to be addressed. Indeed, around the same time that remote ID became a barrier to the FAA's regulatory agenda, the FAA found itself embroiled in litigation over the scope of its authority under Section 336—litigation that would not end well for the agency.

In December 2015, the FAA adopted a rule requiring all UAS operators, including operators of model aircraft governed by Section 336, to register the aircraft with the FAA.[77] The FAA elected to adopt the registration requirement via an interim final rule, which allows federal agencies to circumvent traditional notice-and-comment rulemaking procedures required by the Administrative Procedure Act "when the agency for good cause finds . . . that notice and public procedure thereon are impracticable, unnecessary, or contrary to the public interest."[78] The FAA explained, "Aircraft registration is necessary to ensure personal accountability among all users of the NAS." Furthermore, avoiding the delay of rulemaking procedures was necessary in light of "the current unprecedented proliferation of new [small UAS]," including 800,000 small UAS expected to be sold in the fourth quarter of 2015.[79] Essentially, the FAA circumvented traditional rulemaking procedures to adopt drone registration requirements in advance of the 2015 holiday gift-giving season.

In the interim final rule published in the *Federal Register*, the FAA explained that the registration requirements would apply to model aircraft governed by Section 336. The agency reasoned that it could adopt the requirements notwithstanding Section 336's limitations on FAA jurisdiction because even model aircraft are "aircraft" within the meaning of federal law, and a preexisting provision of the U.S. Code required all aircraft to be registered with the agency.[80]

A model aircraft hobbyist, John Taylor, challenged the interim final rule in the U.S. Court of Appeals for the D.C. Circuit on the grounds that applying the registration requirements to hobbyists violated Section 336 (*Taylor v. Huerta*).[81] The court agreed. Writing for the three-judge panel, then-Circuit Judge Brett Kavanaugh rejected the FAA's argument that it was merely enforcing a preexisting requirement that all aircraft be registered:

> The Registration Rule is a rule that creates a new regulatory regime for model aircraft. The new regulatory regime includes a "new registration process" for online registration of model aircraft. 80 Fed. Reg. at 78,595. The new regulatory regime imposes new requirements—to register, to pay fees, to provide information, and to display identification—on people who previously had no obligation to engage with the FAA. 80 Fed. Reg. at 78,595–96. And the new regulatory regime imposes new penalties—civil and criminal, including prison time—on model aircraft owners who do not comply. See 80 Fed. Reg. at 78,630.[82]

"In short," the court found, "the Registration Rule is a rule regarding model aircraft"[83]—which Section 336 squarely prohibited. Accordingly, the D.C. Circuit invalidated the rule as applied to model aircraft that fall under Section 336.

The D.C. Circuit's decision in *Taylor* came down in May 2017, just a few months after the NPRM concerning flights over people was delayed due to security stakeholder concerns over the need for remote ID. The development presented a new challenge: not only would the FAA need to develop a remote ID solution that would allay the concerns of security stakeholders, but it would need to find a way to require all UAS operators in the NAS to implement the solution even though Section 336 removed a large subset of those aircraft from the agency's regulatory authority. Indeed, despite the schism at the Remote ID and Tracking ARC as to applicability of remote ID to model aircraft, it later became clear that exempting hobbyists and thus cutting out a large proportion of operators (particularly those that on average might be less sophisticated or less inclined to operate safely or follow the rules than their commercial counterparts) was unworkable from a safety and security perspective.

Congress ultimately reinstated the UAS registration rule a few months after the D.C. Circuit invalidated it, again right in time for the holiday season. The National Defense Authorization Act for Fiscal Year 2018, signed into law in December 2017, expressly restored the rules vacated by the D.C. Circuit as they applied to model aircraft.[84] However, the legislation did nothing to address the FAA's underlying authority to regulate hobbyist UAS. To implement remote ID, and likely other future requirements that would emerge as necessary to the safety of the airspace, the FAA would need jurisdiction to regulate hobbyist UAS.

THE INTEGRATION PILOT PROGRAM

Thus, rulemakings to expand Part 107 were delayed indefinitely by the concerns of security stakeholders, and the FAA still lacked statutory authority to adopt broadly applicable remote ID rules. In that climate, the federal government embarked on an initiative that would provide at least some opportunity for expanded UAS operations while also tackling other problems related to the development of UAS policy: the Integration Pilot Program (IPP). Initiated by presidential memorandum in October 2017 and directed to be carried out by DOT and the FAA, the IPP served three objectives: "(i) test and evaluate various models of State, local, and tribal government involvement in the development and enforcement of Federal regulations for UAS operations; (ii) encourage UAS owners and operators to develop and safely test new and innovative UAS concepts of operations; and (iii) inform the development of future Federal guidelines and regulatory decisions on UAS operations nationwide."[85]

The IPP was not the first example of the FAA using public-private partnerships to allow for UAS experimentation and operations that could be conducted out in front of the existing regulatory environment.

In addition to the test sites discussed above that were required by the 2012 NDAA and FMRA, in 2015 the FAA entered into a series of partnerships with specific industry stakeholders to explore certain expanded UAS operations, specifically operations beyond visual line of sight (PrecisionHawk, BNSF Railway) and over people (CNN). Known as the Pathfinder Program, this initiative allowed the partner entities to explore and develop these concepts of operations while providing the FAA with data to inform future rulemakings.[86] The Pathfinder Program was a valuable opportunity for the entities involved to demonstrate the safety of their operations to the FAA; indeed, the Pathfinder participants were the first entities to receive Part 107 waivers for their respective expanded operations.[87]

The success of the Pathfinder Program and the opportunities it provided for the industry participants, coupled with the continued delay of regulations authorizing expanded UAS operations, motivated the industry to advocate for an expanded pilot program in the nature of the IPP. The month before the program was announced, a group of 30 stakeholders representing a wide cross section of UAS industry leaders sent a letter urging the president to initiate "a pilot program that allows state and local governments, along with UAS industry stakeholders, to develop a coordinated effort with the FAA concerning UAS airspace integration." The stakeholders emphasized that such a pilot program would "allow for a data-driven process, within a controlled operational environment, to explore the best options for states and municipalities to address their needs, as it relates to different types of UAS operations."[88]

Consistent with the presidential memorandum, DOT established four objectives for the IPP in a November 2017 *Federal Register* notice: "(1) [t]o accelerate the safe integration of UAS into the national airspace by testing and validating new concepts of beyond visual line of sight operations in a controlled environment, focusing on detect and avoid technologies, command and control links, navigation, weather and human factors; (2) to address ongoing security and safety concerns associated with operating UAS close to people and critical infrastructure and ensure effective communication with law enforcement; (3) to promote innovation in the UAS industry; and (4) to identify the most effective models of balancing local and national interests in UAS integration."[89] Applicants would be limited to state and local governments, but private sector stakeholders could be involved through public-private partnerships.[90]

In May 2018, DOT announced the selection of 10 government entities from a pool of roughly 150 applicants: (a) Choctaw Nation of Oklahoma, Durant, Oklahoma; (b) City of San Diego, California; (c) Virginia Tech–Center for Innovative Technology, Herndon, Virginia; (d) Kansas Department of Transportation, Topeka, Kansas; (e) Lee County Mosquito Control District, Ft. Myers, Florida; (f) Memphis-Shelby County Airport Authority, Memphis, Tennessee; (g) North Carolina Department of Transportation, Raleigh, North Carolina; (h) North Dakota Department of Transportation, Bismarck, North Dakota; (i) City of Reno, Nevada; and (j) University of Alaska-Fairbanks, Fairbanks, Alaska.[91] The winning applications proposed to explore a wide variety of expanded UAS operations, including, among others, BVLOS operations, flights over people, UAS traffic management, medical device delivery, infrastructure inspection, search and rescue and emergency response, and nighttime operations. The scope of operations of the selected applicants was subsequently refined—and in some cases, to the disappointment of the participants, reduced—as the winners worked with the FAA to obtain the requisite authority for the operations. The IPP concluded in October 2020, consistent with the program duration set forth in the presidential memorandum and in subsequent legislation codifying the terms of the program. However, the FAA immediately commenced a successor. Called BEYOND, the follow-on program continued the partnership with eight of the initial IPP participants and identified slightly different, although closely related, objectives: (a) "repeatable, scalable and economically viable" BVLOS operations, with a focus on package delivery, infrastructure inspection, and "public operations"; (b) "[l]everaging industry operations to better analyze and quantify the societal and economic benefits of UAS operations"; and (c) "[f]ocusing on community engagement efforts to collect, analyze and address community concerns."[92]

The IPP served a number of purposes for the UAS industry and its regulatory trajectory. First, for the industry participants partnered with winning government entities, the IPP provided a fast track to securing the

FAA waivers and other regulatory relief needed to conduct the operations. Second, like the Pathfinder Program, the IPP provided (and through its successor will continue to provide) data to the FAA validating the safety of the operations and the risk mitigation techniques employed—although these benefits have yet to be fully realized from an FAA rulemaking perspective. Third, the IPP helped allay concerns among state and local governments interested in regulating drones, an issue which (as discussed later) presents an independent threat to the burgeoning industry. Finally, the program provided a platform for the industry and government alike to demonstrate real-life examples of the beneficial uses of drones, thereby helping improve perception of the technology in the eyes of a skeptical public. The BEYOND program looks set to build on these achievements in an incremental way, though it falls short of the rapid transition to full integration that some thought would follow the IPP.

LAANC: REAL-TIME AIRSPACE AUTHORIZATION AND UAS SERVICE SUPPLIERS

Shortly after the IPP was announced, the FAA made significant progress in the area of controlled airspace authorizations with the deployment of LAANC. As discussed earlier, Part 107 permitted UAS flights only in Class G airspace, requiring a waiver in the form of an airspace authorization to enable operations in the vicinity of airports. The FAA quickly recognized that its existing procedures simply could not keep pace with demand for airspace authorizations. "From September 2016 to July 2017," the FAA reported, "the Agency received 20,566 authorization requests," and as of October 2017, "the Agency ha[d] processed 14,334 and continue[d] to have over 6,000 authorizations in the processing queue." Further, the agency "expect[ed] the queue w[ould] exceed 25,000 pending authorizations within the next 6 months."[93]

In February 2017, the FAA released a concept of operations for LAANC.[94] As envisioned by the FAA, LAANC would enable UAS operators to obtain real-time authorization to operate within the vicinity of airports by submitting an airspace authorization request to an

FAA-approved third-party UAS service supplier (USS). The USS would then use data provided by the FAA (e.g., maps, temporary flight restrictions, and Notices to Airmen) to determine whether the request could be authorized and would send authorization and notification information to both the operator and the FAA.[95] The FAA viewed LAANC as "vital to the safety of the National Airspace System because it would (1) encourage compliance with 14 C.F.R. 107.41 [the regulation requiring ATC approval to operate in airspace other than Class G airspace], (2) reduce distraction of controllers working in the [ATC] Tower, and (3) increase public access and capacity of the system to grant authorizations."[96]

In October 2017, the FAA sought to begin deploying LAANC at a number of airports and air traffic facilities around the country. Similar to the approach it took to expedite the promulgation of rules requiring drone registration, the agency again used an emergency process to clear regulatory hurdles. This time, the FAA sought to circumvent the Paperwork Reduction Act of 1995, which requires agencies to publish in the *Federal Register* and seek public comment on any forms, regulations, or procedures that require the collection of information from regulated entities.[97] The information collection is then approved by OIRA. At the time the FAA sought to deploy LAANC, it already had in place an approved information collection for airspace authorizations for its own waiver process, but it needed to modify the collection to enable LAANC implementation. The FAA sought to take advantage of a rule that allows for emergency processing of information collection requests if the agency can meet certain criteria. Among the possible justifications is demonstrating that it "cannot reasonably comply with the normal clearance procedures" because doing so could result in "public harm."[98] Here, the FAA argued that "the use of normal clearance procedures is reasonably likely to result in further distraction to Air Traffic Controllers and further non-compliant operations," thus creating "pressing safety consideration[s]" that justified circumventing the usual 90-day procedure.[99] OIRA granted the request in seven days, enabling the FAA to deploy LAANC.[100]

The agency began with what it called a "prototype evaluation" at a series of air traffic facilities around the country, allowing the first two USS entities that received FAA authorization—AirMap and Skyward—to provide LAANC services.[101] In April 2018, the FAA moved to a nationwide beta test that rolled out LAANC capabilities incrementally at roughly 300 air traffic facilities that covered 500 airports nationwide by September 2018.[102] In November 2018, the agency announced that LAANC had processed more than 50,000 applications for airspace authorization and had been expanded to include 14 service suppliers.[103] As of December 2019, those figures had increased to 170,000 and 21, respectively.[104]

LAANC was an important development in the UAS regulatory space not just because it solved a pressing problem but also because it demonstrated the feasibility of the third-party service supplier model. Such suppliers are likely to be an integral part of future initiatives to enable full integration of UAS into the national airspace, most notably UAS traffic management (UTM). It has long been recognized that full UAS integration will create a complex airspace at low altitudes, including routine BVLOS UAS operations and simultaneous operation of numerous aircraft. Ensuring safety in this complex airspace will require a traffic management system akin to the air traffic control system used for manned aircraft. NASA has been collaborating with the FAA since 2014 to determine the requirements and assess available technologies for a UTM system.[105] USSs will likely provide the infrastructure and technological solutions necessary to enable the UTM system, and the FAA's experience with LAANC will inform the agency's development of that system.[106]

FAA REAUTHORIZATION ACT OF 2018

Throughout 2018, the FAA engaged in a number of efforts aimed at expanding UAS integration, including enabling expanded operations through waivers, getting the IPP off the ground, and rolling out LAANC at airports across the country. Conspicuously absent from the FAA's activities was rulemaking to allow expanded UAS operations beyond Part 107 and continue the FAA's regulatory trajectory toward full integration. Indeed, as of October 2018, the FAA's integration chart had evolved substantially to include both regulatory and airspace management components (see Figure 2).

However, remote ID continued to be the key to all future regulatory development, and no apparent path was in sight for implementing a

FIGURE 2
Path to UAS integration

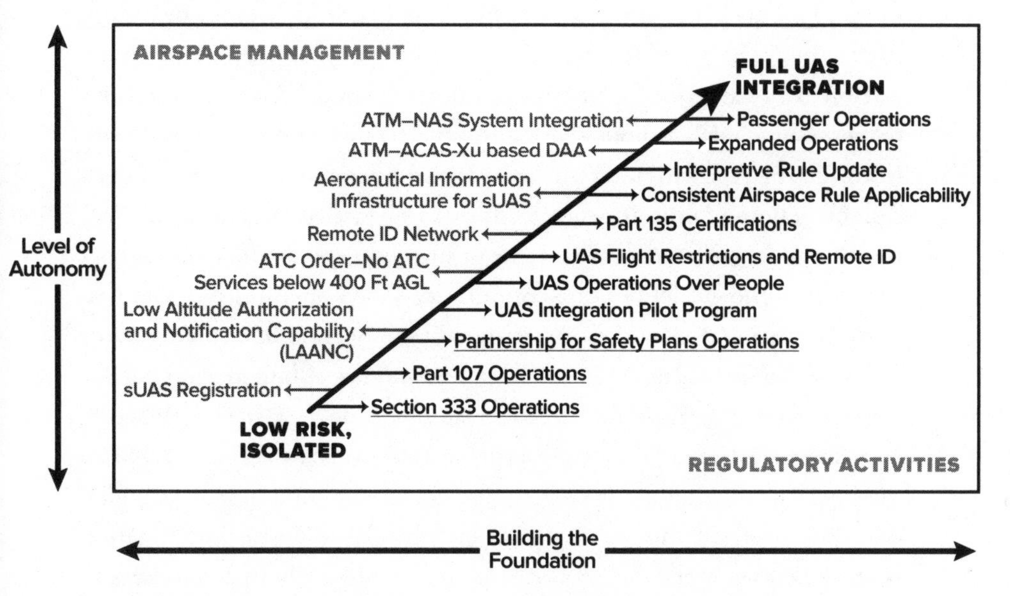

Source: DJI Enterprise, "DJI FlySafe—AirWorks 2018—Policy Panel," November 30, 2018, YouTube video, 1:01:14, https://www.youtube.com/watch?v=QdcXQxgCMfA.
Notes: Figure reproduced from original. ACAS-Xu = Airborne Collision Avoidance System X; AGL = above ground limit; ATC = air traffic control; ATM = air traffic management; DAA = detect and avoid; NAS = National Airspace System; sUAS = small unmanned aerial system; UAS = unmanned aircraft system.

remote ID solution and resolving the concerns raised by security stakeholders, at least in part because of the FAA's apparent lack of authority over Section 336 hobbyists.

Congress solved that problem by passing the FAA Reauthorization Act of 2018.[107] Signed into law in November 2018, the act reauthorized the FAA for a period of five years and included provisions covering a wide variety of aviation issues, including commercial airline passenger protection, airport infrastructure, flight attendant working conditions, and aircraft noise. Unmanned aircraft were a major focus of the act, with numerous provisions requiring new reports, studies, or agency action relating to a wide variety of UAS policy issues. Most important for the UAS industry, the act finally addressed the Section 336 hobbyist loophole.

Section 349 of the Reauthorization Act repealed Section 336 of the FMRA, replacing it with a provision that gave the FAA significantly more regulatory authority over hobbyist operators as well as flexibility in deciding what operators fall under that category. Section 349 retained a regulatory carveout for hobbyists, allowing model aircraft operations to be conducted "without specific certification or operating authority from the [FAA]," provided the aircraft and operators meet certain conditions. Those conditions dictate, for example, that (like Section 336) operations remain within visual line of sight and are conducted in accordance with the guidelines of a community-based organization, and that (new in Section 349, and more stringent than Section 336) the operator has passed an aeronautical knowledge test and has obtained prior authorization from the FAA for any operations outside Class G airspace.[108] Moreover, Section 349 not only expressly authorized the FAA to update the operational parameters of the hobbyist exemption but also provided that "[n]othing in this section prohibits the [FAA] from promulgating rules generally applicable to unmanned aircraft, including [model aircraft covered by Section 349]," related to registration and marking, remote ID, and "other standards consistent with maintaining the safety and security of the national airspace system."[109] Accordingly, the act gave the FAA the requisite authority to push ahead with broadly applicable remote ID rules, which would then enable the agency to expand commercial operations under Part 107.

The Reauthorization Act included numerous provisions related to UAS. These provisions imposed deliverables on the FAA, DOT, the Government Accountability Office, and other federal agencies, and pertained to areas such as airspace management,[110] the Part 107 waiver process,[111] regulatory compliance and enforcement against unlawful operations,[112] safety standards,[113] the protection of critical infrastructure,[114] test programs,[115] package delivery,[116] privacy,[117] public and emergency operations,[118] state and local roles in UAS regulation,[119] spectrum,[120] regulatory fee mechanisms,[121] and UAS integration generally.[122]

The act also took significant steps to mitigate potential UAS threats, such as expanding counter-UAS authority (i.e., the authority to "detect, identify, monitor, and track," "warn the operator of," "disrupt control of," "seize or exercise control of," "seize or otherwise confiscate," or "use reasonable force, if necessary, to disable, damage, or destroy" a UAS) to two additional federal agencies: the Department of Homeland Security and the Department of Justice.[123] As the Department of Energy and the Department of Defense had received such authority in prior legislation,[124] the Reauthorization Act brought the number of federal agencies with counter-UAS authority to four. The act also addressed the potential dangers of UAS operations near airports—a timely provision given that reports of unauthorized UAS flying near London's Gatwick Airport would grab worldwide attention mere weeks later.[125] The act emphasized that "the unauthorized operation of [UAS] near airports presents a serious hazard to aviation safety;"[126] required the FAA to deploy counter-UAS systems at five airports;[127] and added a new provision to the U.S. criminal code, criminalizing knowingly or recklessly interfering with manned aircraft operations using a UAS and knowingly operating a UAS in close proximity to airport runways.[128]

PART 135 OPERATIONS: THE EMERGENCE OF UAS USE OF EXISTING REGULATORY PROCESSES

The year following reauthorization, 2019, was notable for the UAS industry because it brought a change in regulatory perspective toward using existing regulatory processes to enable expanded UAS operations. Until then, the UAS policy discussion largely focused on the ways in which UAS are different from manned aircraft and pose unique regulatory challenges. Consistent with that perspective, much of the conversation centered around how to circumvent or obtain relief from existing requirements, such as the statutory airworthiness certification requirement or ill-fitting safety regulations designed for manned aircraft.

But 2019 saw a distinct shift in the policy discourse toward the ways in which UAS—and scalable commercial UAS operations—function similarly to manned aircraft, and, accordingly, how UAS operators could take advantage of existing procedures to obtain FAA operational authorization. The delay in UAS-specific rulemakings undoubtedly contributed to this shifting dialogue, but the increasing sophistication of the drone industry was another relevant factor. Years of robust testing and development of UAS technologies and navigating FAA regulatory procedures had resulted in some operators and platforms capable of withstanding the level of scrutiny that the FAA places on manned aircraft operations.

This shift manifested most directly in the context of UAS package delivery. As discussed above, Part 107 allowed package delivery operations only on an intrastate basis. Yet the standing prohibition on using UAS for interstate or international transportation could still apply to an operator that uses wholly intrastate legs of delivery if the operator's delivery network as a whole is interstate. Thus, Part 107 remained unavailable for the most talked-about drone delivery use case—last-mile package delivery by nationwide delivery services. However, the procedures used by manned aircraft air carriers remained available. As previously discussed, under that framework air carriers are statutorily required to hold three types of certifications: (a) an operating certificate from the FAA; (b) an airworthiness certificate from the FAA; and (c) economic authority from DOT.

In 2019, two industry players broke new ground as the first UAS operators to obtain authorizations from the FAA to conduct operations under Part 135 of the agency's rules, which governs air carriers.[129] Wing Aviation, a subsidiary of Google's parent company Alphabet, obtained the nation's first Part 135 certificate for UAS air carrier operations, securing a Single Pilot certificate.[130] Single Pilot certificates enable operations only by the operator named in the certificate and only for one drone flight at a time. In September 2019, UPS Flight Forward obtained a Standard air carrier certificate, allowing it to operate on a larger scale with an unlimited number of operators, aircraft, and routes.[131] In both cases, the FAA also used the Section 333 exemption process (although the appropriate term for this relief is now a Section 44807 exemption, as the Reauthorization Act had replaced Section 333 of the FMRA with a substantially similar provision and codified it in the U.S. Code[132]), providing each operator with relief from certain regulations "to conduct air carrier operations in accordance with part 135, upon receipt of . . . [an] air carrier certificate from the [FAA]."[133] The agency issued separate exemptions to authorize operation of the aircraft without airworthiness certificates, but it explained that the applicants had applied for type certificates for their aircraft and that airworthiness certificates would supersede the exemptions once the type certification process was complete. In fact, the exemptions specified that "continued operations under that exemption are predicated on continued progress toward that airworthiness certificate."[134]

Another notable fact about the Wing and UPS experiences obtaining Part 135 authorization is that both applicants were participants in the IPP. Wing tested commercial package delivery in conjunction with the Mid-Atlantic Aviation Partnership at Virginia Tech,[135] and UPS deployed a pilot program for delivering medical samples at WakeMed hospital in Raleigh, North Carolina, in collaboration with UAS platform provider Matternet.[136] In both cases, the FAA explained that the exemptions authorizing Part 135 operations were being issued "in connection with" the IPP, and that the applicants' participation in the program contributed to the agency's determination that granting the exemptions was in the public interest.[137] The FAA's treatment of these IPP participants thus demonstrates the value of participation in FAA-sponsored test and pilot programs not just for obtaining Part 107 waivers but for securing authority for increasingly complex operations.

DOT also has taken steps to facilitate UAS package delivery in its role as the agency that grants economic authority for air carrier operations. DOT already had in place a process for exempting a subset of air carriers, called air taxi operators, from the economic authority requirement if the carriers meet a series of requirements; for example, they must hold liability insurance and not use large aircraft (defined as aircraft with a payload capacity of more than 18,000 pounds). Further, the department had recognized that this air taxi designation and exemption process—governed by Part 298 of DOT's regulations—could be a good fit for UAS operators. In April 2018, the agency released a *Federal Register* notice explaining that "[f]or UAS operators looking to transport goods for compensation, an exemption under part 298 is an appropriate form of economic authority," and that "[t]he Department [would] consider whether granting the exemption is appropriate based on the specific facts and circumstances of each proposed operation."[138]

THE STATE AND LOCAL CHALLENGE

Under federal law, "[t]he United States Government has exclusive sovereignty of airspace of the United States" and U.S. citizens have "a public right of transit through the navigable airspace."[139] Federal law also vests in the FAA the authority to regulate air navigation and aviation safety.[140] Yet the FAA's pace in adopting regulations governing UAS and the lack of clarity surrounding how traditional areas of local police power would apply to this new aviation technology have bred state and local attempts to regulate UAS. These developments have raised important questions about the scope of federal preemption in the aviation space.

In 2015, the FAA Office of the Chief Counsel issued a guidance document describing the scope of federal preemption over state and local UAS regulations.[141] The guidance document explained both that "[s]ubstantial air safety issues are raised when state or local governments attempt to regulate the operation or flight of aircraft" and that "[a] navigable airspace free from inconsistent state and local restrictions is essential to the maintenance of a safe and sound air transportation system."[142] Further, it identified the types of local restrictions that would pose such a threat, including "[o]perational UAS restrictions on flight altitude, flight paths[,] operational bans[,]" and rules that "mandat[e] equipment or training for UAS related to aviation safety such as geo-fencing."[143] At the same time, the guidance document acknowledged that regulation affecting UAS in areas "traditionally related to state and local police power—including land use, zoning, privacy, trespass, and law enforcement operations"—may be permissible notwithstanding federal primacy over the airspace.[144]

Despite the long-standing exclusive federal role in aviation regulation and the FAA's attempts to clarify that role in the context of UAS, some state and local governments forged ahead with legislative enactments regulating all manner of UAS operations in ways that appeared to infringe upon the federal government's occupation of the field of air navigation, conflict with federal regulations, or both. For instance, in December 2016, the City of Newton, Massachusetts, passed an ordinance prohibiting the operation of UAS over private property at an altitude of less than 400 feet without the express permission of the property owner, over Newton city property without permission from the city, and beyond the visual line of sight of the operator.[145] The ordinance also required all UAS operators to register the aircraft with the city.[146] A resident of the city challenged the law, and the U.S. District Court for the District of Massachusetts held that all three provisions were preempted by federal law because they impermissibly conflicted with the FAA's regulation of navigable airspace, UAS integration, and aviation safety.[147]

By contrast, in late 2020, the U.S. District Court for the Western District of Texas rejected a preemption challenge by the National Press Photographers Association to a state law that restricted drone flights over certain types of state-defined critical infrastructure. The court found that these relatively narrower restrictions on UAS flights did not raise the same conflicts with federal law as had the City of Newton's ordinance.[148] While the court granted the state's motion to dismiss the preemption challenge to the infrastructure overflight provisions, it allowed to proceed the plaintiffs' First Amendment claims, challenging a different aspect of the law that criminalized image capture via UAS in certain contexts and exempted some industries, but not newsgatherers, from its purview.[149] If ultimately successful, these First Amendment claims could lead to significant limitations on governmental authority to restrict UAS use in certain contexts.

Unmanned aircraft also have been the subject of study by the Uniform Law Commission (ULC), a national organization that develops model state laws. The ULC established a committee to draft a statute on tort law related to UAS, focused primarily on whether state trespass laws need to be updated to take into account this new type of aircraft. The committee's early draft would have imposed a per se rule for aerial trespass using a UAS, under which a UAS operator would be liable for trespass simply for operating within 200 feet above private property without the owner's permission.[150] The proposal disregarded the state of the law on aerial trespass, which requires *both* intrusion into the "immediate reaches" of the airspace above a person's property *and* for that intrusion to "interfere[] substantially with the other's use and enjoyment" of the property.[151] Indeed, in a joint letter from each agency's general counsel, the FAA and DOT weighed in on the draft, clarifying that the proposal exceeded the permissible role of state and local regulation as laid out in the FAA's guidance document, and asserting that the proposed per se rule "would be in tension with decades of established precedent in the Federal courts, which have rejected the notion of applying the traditional elements of basic trespass law to aircraft overflight of private property."[152] After significant discussions with stakeholders from industry, think tanks, and other organizations, the committee reached a compromise; the revised proposal identified a series of enumerated, UAS-specific factors for courts to weigh when considering whether a UAS has committed an aerial trespass.[153] The committee presented the draft for a vote at the ULC's annual meeting in 2019, but it was ultimately rejected by the full commission over last-minute concerns raised by property rights groups such as the Joint Editorial Board for Uniform Real Property Acts, the property section of the American Bar Association, and the American College of Real Estate Lawyers.[154] The ULC appears to have abandoned further efforts in this area for now and has shelved its uniform law proposal. As the ULC process concluded, the American Law Institute (ALI) began considering similar issues regarding UAS and trespass in revising its Restatement of Property—a process that likewise sparked concern among UAS stakeholders. Although originally set for a vote in 2021, ALI delayed consideration of the revision until at least 2022.

The FAA is expected to take additional action related to the balance of authority between federal, state, and local governments, including revising the 2015 guidance document to further clarify the boundaries of state and local authority over UAS regulation.[155] In addition, as previously discussed, one of the stated goals of the IPP was to "test and evaluate various models of State, local, and tribal government involvement in the development and enforcement of Federal regulations for UAS operations." Although the FAA made great strides through the IPP in areas such as the testing and deployment of novel operations involving both state and local governments and industry stakeholders, the agency made less progress on determining the proper legal and enforcement roles of the competing sovereigns. And the BEYOND program that replaced the IPP dropped that objective altogether. Whatever steps the agency takes, the balance between federal and state authority will be an ongoing challenge, particularly if the FAA continues to experience delays in its regulatory agenda and negative public perception of drones continues to prompt action by state and local legislators.

ON THE HORIZON: REMOTE ID, EXPANDED OPERATIONS, SECTION 2209, UAS TRAFFIC MANAGEMENT, AND SPECTRUM

What will be the defining features of UAS regulation in 2021—and the years to come? The UAS industry currently stands at a regulatory inflection point. The growing pains of the past several years have largely subsided, leading to a regulatory environment in which the FAA and other federal agencies are poised to undertake a number of significant actions in quick succession—actions that should transform UAS integration from a distant aspiration articulated in a legislative provision to a palpable reality. The areas to watch include (a) remote ID, (b) expanded UAS operations through FAA rulemaking, (c) protection of critical infrastructure pursuant to Section 2209 of the 2016 Extension Act, (d) UAS traffic management, and (e) spectrum policy. Each of these areas is discussed briefly in turn.

REMOTE ID

Implementation of remote ID is the key to future enabling regulations for commercial UAS operations. After numerous delays, the FAA finally released an NPRM proposing remote ID regulations in late December 2019. The NPRM proposed to establish two categories of remote ID capabilities: standard and limited. Standard remote ID would publish information about the aircraft and control station (the remote ID message) using both radio frequency broadcasting using unlicensed spectrum (broadcast remote ID) and transmission via the internet to a remote ID USS (network remote ID); UAS equipped with standard remote ID would be permitted to operate BVLOS (with the proper FAA authorization, of course, as Part 107 currently prohibits BVLOS operations without a waiver).[156] Limited remote ID would publish the remote ID message only via internet connection to a USS; the UAV would need to remain within the visual line of sight (specifically, no more than 400 feet from the control station).[157] For either category of remote ID capability, the remote ID message would contain similar information: (a) the UAS ID (serial number of UAS or session ID); (b) latitude/longitude and barometric pressure altitude of the control station (and, for standard remote ID UAS, of the aircraft as well); (c) time stamp; and (d) emergency status of the UAS.[158] UAS not equipped with remote ID generally would be permitted to operate only in what the NPRM called "FAA-recognized identification areas" (FRIAs), which would be limited areas intended to accommodate primarily recreational or research use.

The remote ID proposal in the NPRM was ambitious and was intended to serve as a building block for the broader UAS Traffic Management system. Requiring remote ID signals to feed into a network would allow for a clearer picture of where unmanned aircraft were flying and allow UAS Service Suppliers to develop systems that would eventually allow something like automated traffic control for unmanned aircraft. This proposal met with some controversy, with operators raising questions about affordability of network connections, the ability to fly in places without internet access, and privacy and storage of flight data.

A year later, in December 2020, the agency promulgated final rules in the remote ID proceeding that substantially cut back on the ambition of the proposed approach. Rather than attempt to build a networked system that would serve as a platform for UAS Traffic Management, the FAA elected to focus on solving the immediate challenge of allowing safety and security entities to identify aircraft at close range. Citing "challenges" that "it had not foreseen or accounted for" with respect to network remote ID, the FAA opted for a "simplified approach" of using only broadcast remote ID[159]—in other words, UAS will be required to broadcast certain information using short-range, unlicensed spectrum, and will not have to connect to a network in order to transmit this information to a central UAS service supplier. The agency has not abandoned completely the broader vision of a fully networked UAS ecosystem, but it has kicked that can down the road, in much the same way it did with interstate air carriage and Part 107. While this can be seen as a victory for those who wanted easier, cheaper solutions for remote ID, those who were hoping that remote ID would represent a quantum leap toward greater capability and widely expanded operations were disappointed.

Even the FAA's simplified remote ID approach has complexity, particularly to accommodate the millions of preexisting UAS that need to be retrofitted with remote ID capabilities. Under the remote ID rules, with limited exceptions, UAS that are required to be registered with the FAA (all UAS above 0.55 pounds, other than certain aircraft of the federal government) must fall into one of three categories to operate in U.S. airspace: (1) the UAS is manufactured with broadcast remote ID capabilities (called standard remote ID—not to be confused with the network- plus broadcast-based standard remote ID from the NPRM); (2) the UAS was manufactured before the remote ID compliance deadline for UAS manufacturers and is outfitted with a "broadcast module" that enables the aircraft to broadcast the remote ID message; or (3) the UAS is operated at a FRIA.[160] Standard remote ID UAS are required to broadcast both the location of the control station and the location of the aircraft as part of the remote ID message and are permitted to operate BVLOS with the requisite FAA authorization. UAS equipped with broadcast modules need only transmit the takeoff location in lieu of the real-time location of the control station, but because this makes the operator harder to locate, they must remain within visual line of sight. FRIA operations also must be conducted within visual line of sight.[161] Both standard remote ID UAS and broadcast modules must be produced in accordance with FAA-accepted means of compliance demonstrating that the aircraft or module meets the FAA's remote ID performance-based requirements such as the accuracy of location information, message frequency, and tamper resistance.[162]

The final remote ID rules were published in the Federal Register on January 15, 2021, starting a 60-day clock for the rules to become effective.[163] The FAA stayed the effective date of the rules pursuant to a White House memorandum to allow the incoming administration to review new and pending rules, and the rules became effective April 21, 2021.[164] Under the rules, by September 2022 manufacturers may produce only standard remote ID UAS. Twelve months after that—September 2023—

UAS operators will be required to use remote-ID-equipped UAS to operate outside a FRIA in U.S. airspace.[165] Although one of the FAA's motivations for adopting the simpler remote ID rules was to ease implementation, the fact that the rules will not take full effect until 2023 underlines the long lead times associated with rulemaking procedures, and once again underscores that further development such as expanded operations and UAS Traffic Management likewise will take years to implement. Moreover, even the FAA's simplified approach to remote ID has raised some implementation questions, including whether limiting broadcast remote ID to unlicensed spectrum was unnecessarily prescriptive, and why the agency's otherwise "performance-based" approach does not allow for broadcast modules to be engineered to comply with "standard" remote ID protocols and thus allow BVLOS operation by retrofitted UAS.

EXPANDED OPERATIONS

Because remote ID was the hurdle preventing the FAA from revising Part 107 to allow routine expanded operations, adoption of those final rules allows the FAA to focus on that objective. The agency got a jump-start on this process by adopting final rules on flights over people and nighttime operations simultaneously with the final remote ID rules in December 2020.[166]

The final rules largely track an NPRM that the FAA published in February 2019 proposing rule changes to enable these expanded operations. Much like the NPRM that was never published in 2017 was expected to do, the FAA's approach to operations over people draws from the recommendations of the Micro UAS ARC. Specifically, the FAA adopted a multi-tier framework based on the relative level of risk posed by the aircraft. Category 1 includes all UAS weighing 0.55 pounds or less, which can be operated over people under Part 107 as long as the aircraft does not have any exposed rotating parts that would lacerate human skin on impact.[167] Category 2 includes UAS weighing more than 0.55 pounds, which are permitted to operate over people if they meet three requirements: (a) the aircraft, upon impact with a person, "will not cause injury to a human being that is equivalent to or greater than the severity of injury caused by a transfer of 11 foot-pounds of kinetic energy upon impact from a rigid object"; (b) the aircraft does not have exposed rotating parts that could lacerate human skin; and (c) the aircraft does not have an FAA-identified safety defect.[168] Category 1 and 2 UAS can operate over open-air assemblies of people only if they are remote-ID-compliant (either standard or using broadcast modules).[169]

Category 3 includes UAS meeting the same design criteria as Category 2 except with the injury threshold at 25 foot-pounds rather than 11.[170] Because of the increased risk posed by Category 3 UAS, operators of such aircraft are subject to additional operational constraints. Specifically, Category 3 UAS operators are prohibited from operating over open-air assemblies of people regardless of whether they have remote ID, and their operations must be conducted either at a closed- or restricted-access site or without maintaining sustained flight over nonparticipating people.[171] Consistent with the approach taken for remote ID, UAS in Categories 2 and 3 must be produced in accordance with a Federal Communications Commission (FCC)-accepted means of compliance that demonstrates compliance with the design standards.[172] Finally, in recognition of the increasing ability of UAS to use regulatory structures initially designed for manned aircraft, Category 4 includes UAS that have an airworthiness certificate, which are permitted to operate over people in accordance with the operation limitations specified in the approved Flight Manual.[173] Like UAS in the first two categories, Category 4 UAS can operate over open-air assemblies of people only if the aircraft is remote-ID-compliant.

The FAA adopted this multi-tier framework over the objections of industry stakeholders, who asserted in comments in response to the NPRM both that it would be unduly onerous to develop rigorous testing methods to demonstrate the level of injury that would be caused by the UAS in the event of a collision as compared with impacts by a rigid object, and that the particular injury thresholds chosen for Categories 2 and 3 were too conservative.[174] Stakeholder response to the NPRM made a notable impact in one area, however: while the NPRM proposed to prohibit all operations over people—which industry commenters asserted would drastically reduce the utility of the expanded regulations[175]—the final rules allow UAS in the first three categories to operate over moving vehicles under the same restrictions imposed on all Category 3 UAS operations: either at a closed site or without maintaining sustained flight over the vehicles.[176] Category 4 UAS can operate over moving vehicles consistent with the approved Flight Manual.[177]

The rules authorizing routine Part 107 operations at night are much simpler, and borrow from the FAA's authorized thousands of operators to pilot drones for commercial purposes under the Part 107 waiver process. The final rules allow UAS operations at night as long as the aircraft is equipped with sufficient anti-collision lighting and the pilot has taken either an initial knowledge test or recurrent training that contains a nighttime component.[178] Flights at night can commence once the rule becomes effective (60 days after publication), as can flights over people for compliant Category 1 aircraft. For the other categories of aircraft that may operate over people, the FAA anticipates as much as a year of lead time given the need to process the means of compliance submitted by manufacturers (for Categories 2 and 3) and issue appropriate airworthiness certificates (for Category 4).[179]

In addition to remote ID and operations over people and at night, the FAA is pursuing rulemaking in other areas to enable expanded UAS operations. Contemporaneously with the NPRM on flights over people and at night published in February 2019, the FAA released an Advance Notice of Proposed Rulemaking (ANPRM) on "Safe and Secure Operations of Small [UAS]."[180] The ANPRM seeks comments on issues likely to inform future rulemakings, including stand-off distances,[181] payload restrictions,[182] critical system design requirements,[183] performance limitations,[184] and (as discussed later) UAS Traffic Management operations.[185] Another major FAA priority for future rulemaking is enabling routine BVLOS flights, although the feasibility of such regulations depends not just on remote ID but also likely on the implementation of UAS traffic management.

Finally, new UAS-specific regulations for air carriers also may be on the horizon: the FAA Reauthorization Act of 2018 directs the FAA to "update existing regulations to authorize the carriage of property by operators of small [UAS] for compensation or hire."[186] The same provision also authorizes DOT to revise its regulations to establish blanket economic authority for small UAS air carriers. Of course, the final step in the FAA's regulatory trajectory for UAS integration is not package delivery by small UAS but advanced air mobility (AAM)—a broader, more advanced concept of operations that includes the carriage of people, large-scale cargo delivery, and the provision of public services by large unmanned aircraft. Urban air mobility, one AAM use case, is touted for its potential to improve transportation systems and decrease ground congestion in urban and suburban areas by allowing short flights along predesignated routes, using building rooftops as appropriate for takeoff and landing. Given the human lives at stake and increased risk posed by larger aircraft, the regulatory regime for AAM will probably look more like those governing manned aircraft than those governing UAS. In early 2021, FAA officials indicated that the agency was working with more than 30 companies on AAM aircraft, and that type certifications for some AAM aircraft could be issued within the year.[187]

SECTION 2209

The FAA is also expected to take steps in the medium term to implement a process to protect critical infrastructure from UAS overflights. Section 2209 of the 2016 FAA Extension Act required the FAA to develop a process that would allow "applicants" to petition the FAA to "prohibit or restrict" UAS operations in "close proximity" to a "fixed site facility."[188] The shorthand for this in the industry is a procedure for restricting flights near or over "critical infrastructure," but the actual statutory language is broader, applying to "critical infrastructure such as energy production, transmission, distribution facilities and equipment, and railroads" (the last of which was added in 2018 by the Reauthorization Act), as well as "oil refineries and chemical facilities," "amusement parks," and "other locations that warrant such restrictions."[189] The agency has yet to implement the process required by Section 2209 and has missed both the original 180-day deadline from the 2016 act and the March 31, 2019, deadline imposed by the 2018 Reauthorization Act. As of the Fall 2020 Regulatory Agenda, the anticipated date for an NPRM had slipped to October 2021.[190]

Once the regulations are in place, the Section 2209 process is expected to have a positive impact on allaying concerns of state and local governments, several of which have taken measures into their own hands to protect critical infrastructure in the absence of FAA regulations or procedures.[191] Although some states' critical infrastructure laws include a sunset provision that terminates the state law once the FAA's procedure is in place[192]—and at least one such law is currently being challenged in federal court[193]—these laws present the same concerns as other UAS operational bans and restrictions given their interference with federal sovereignty over the airspace. A centralized federal procedure will obviate the need for critical infrastructure regulation at the state and local level, leading to a safer and more efficient regulation of the airspace.

UAS TRAFFIC MANAGEMENT

LAANC is now operational at hundreds of airports across the country, and the remote ID rules are in effect. Thus, UTM is the clear next step in airspace management for the UAS industry. NASA and the FAA are collaborating on developing technologies and processes that will underpin the UTM system. In addition, the FAA has established the UTM Pilot Program (UPP) and selected three test sites to test the UTM concept. The FAA has explained the following:

> UTM services to be demonstrated in the UPP include sharing of flight intent between operators, the ability for a UAS Service Supplier . . . to generate a UAS Volume Reservation . . . —a capability providing authorized USSs the ability to issue notifications to UAS Operators regarding air or ground activities relevant to their safe operation—and share it with stakeholders.[194]

The pilot program is being implemented in accordance with Section 2208 of the 2016 FAA Extension Act, which directed the FAA to establish a research plan for UTM as well as a two-year pilot program for UTM implementation in coordination with NASA, the Drone Advisory Committee, the FAA's research advisory committee, and representatives of the UAS industry.[195] Section 376 of the 2018 Reauthorization Act builds on this pilot program and requires the FAA, "in conjunction with completing the requirements of section 2208," to develop a UTM implementation plan for services "that expand operations beyond visual line of sight, have full operational capability, and ensure the safety and security of all aircraft."[196] The plan must be completed within one year after the UPP concludes.

In addition, the 2018 act broadened the UPP to (a) cover testing of increased operations and density over test ranges and "other sites determined by the Administrator to be suitable for UTM testing," including IPP locations; (b) permit testing of remote ID technologies; and (c) permit blanket waivers for UTM pilot program selectees, rather than the case-by-case waivers required under Part 107.[197] The act also imposed a number of requirements on the long-term implementation plan for UTM, such as developing safety standards and outlining the roles of industry and government with particular attention to the model used for LAANC; and the act required the FAA to take into account additional considerations, including remote ID of cooperative and noncooperative aircraft, coordination with air traffic control, and cybersecurity.[198] The act further directs the administrator to establish expedited procedures for approval of UTM services in low-risk areas.[199]

The 2018 act also sought to expedite UTM deployment, directing the FAA to determine whether certain UTM services can operate safely prior to the completion of the required plan and, if so, to establish requirements for safe operation of these services.[200] The administrator is required to "provide expedited procedures" for making the determination in areas where the operation of unmanned aircraft poses a low risk, including croplands and other noncongested areas.[201] Given that the remote ID NPRM proposed the network-based USS model in part because the FAA envisions remote ID as a "building block" for UTM, the agency's decision in the final rule to require only broadcast remote ID using unlicensed spectrum raises questions about the agency's trajectory for UTM implementation.

SPECTRUM

Ensuring that the UAS industry has sufficient access to radio frequency spectrum is a separate regulatory issue that will require considerable attention in the coming years. Currently, UAS command and control and payload communications are transmitted primarily using unlicensed spectrum—a system that is unworkable in the long term both because of the limited capacity of the spectrum and because unlicensed users do not enjoy interference protection from other users.

Multiple potential solutions can provide the UAS industry with sufficient spectrum to meet its growing needs as UAS proliferate and the FAA continues to allow expanded operations. This UAS spectrum policy debate is taking place among stakeholders at the FAA, at the Federal Communications Commission (FCC), and within working groups of the FCC's Technological Advisory Council[202] and the National Telecommunications and Information Administration's (NTIA) Commerce Spectrum Management Advisory Committee.[203] The options being discussed include (a) internationally allocated aeronautical spectrum, including 5030–5091 MHz (the C-band) and 960–1164 MHz (the L-band); (b) terrestrial wireless spectrum used by commercial mobile radio service providers (in other words, cellular carriers), including that used for fifth-generation wireless services (5G); and (c) satellite spectrum, specifically the Ku- (12–18 GHz) and Ka-bands (17–30 GHz). Given that each of these frequency bands and their associated regulatory regimes have benefits and drawbacks, the industry ultimately will likely use some combination of these options—as well as some unlicensed spectrum.

Solving the UAS spectrum problem will require significant interagency coordination, including the FCC, FAA, and NTIA (which manages spectrum allocated for federal use). The 2018 Reauthorization Act tasked these three agencies with developing a report on the feasibility of using C-band and L-band aeronautical spectrum for UAS operations.[204] After the FAA and FCC each solicited public comment in parallel proceedings that were said to be the product of interagency tension, the FCC submitted the report to Congress in August 2020. The report expressed support for using the C-band to support UAS operations and recommended that the FCC initiate a rulemaking proceeding to adopt the necessary service and licensing rules.[205] With respect to the L-band, citing "challenges" raised by the "extensive" incumbent use of the band, the report declined to recommend that the FCC move forward with a proceeding to enable UAS use of L-band spectrum.[206] Finally, the report noted that alternative frequencies currently licensed under flexible use service rules could be a promising option for UAS communications, particularly for BVLOS and other network-based use cases.

CONCLUSION

The regulatory journey of the domestic UAS industry over the past decade has provided significant insights into how to integrate a new technology into an already heavily regulated field. Although the pace has been relatively slow, the progress has been significant, and the lessons have been numerous. The UAS industry's experience has demonstrated how a purportedly narrow legislative carveout—such as the one for model airplane hobbyists—can hamper regulatory progress for an entire industry, as can concerns held by federal agencies that have relevant interests but lack direct regulatory authority. The experience also has illustrated the extent to which the regulatory needs of an industry, and the ability to navigate more complex regulatory processes, can evolve with the technology and the sophistication of the regulated entities.

For its part, the FAA has been successful in making its policy development and rulemaking as efficient as possible despite the larger obstacles it has faced. For example, it has leveraged stakeholder expertise through ARCs; used third-party service suppliers; used the experience of early movers authorized through test or pilot programs, waivers, or exemptions to inform and validate subsequent policies; and taken advantage of emergency regulatory procedures. Of course, the availability of regulatory relief through test programs, pilot programs, and case-by-case determinations creates a somewhat uneven playing field for operators, giving the more sophisticated and well-connected entities a clearer path to early mover status or one-off authorization.

The future success of the UAS industry depends on the FAA's ability to continue to leverage the tactics that have worked well, while minimizing the likelihood of large obstacles that cause multiyear delays. With the past decade behind it, the agency appears poised to do just that.

CHAPTER TWO

WHO WANTS A DRONE ANYWAY? THE LAW DEVELOPS TO ACCOMMODATE THE PROMISE OF COMMERCIAL DRONES

GREGORY S. WALDEN

Who wants a drone anyway? A quick answer is that millions of people around the world, including hundreds of thousands of Americans, own and operate drones.[1] Consider the numbers: as of March 2021, the Federal Aviation Administration (FAA) had registered 873,540 drones. Of them, 367,848 were registered by commercial operators and 502,105 by recreational users. For commercial operations authorized under FAA's Part 107 rules,[2] 223,634 pilots had obtained a remote pilot certificate with unmanned aircraft system (UAS) rating.[3]

Each of these numbers is an undercount. With respect to the number of drones, many are not registered—despite the small fee ($5.00) and the easy online registration process. The magnitude of this undercount can't be determined. Neither manufacturers nor retailers are required to provide sales numbers, and apparently no one does. With respect to pilots, a sizable number of commercial operators likely have not obtained a pilot certificate given that a person must travel to an FAA-approved testing center and pay a fee to take the test. Recreational pilots are not required to obtain a certificate; they may well be operating unregistered drones in addition to flying them without a certificate.

Who are these more than one million Americans? FAA records do not provide any breakdown by age or occupation. Let's ask instead for what purposes are drones made, purchased, and operated?

DEVELOPMENT OF DRONE OPERATIONS

At the dawn of the 20th century, engineers and weekend tinkerers designed early aircraft models; they regarded these smaller, lighter, and pilotless aircraft more as toys for enjoyment than for business purposes. Modelers established the Academy of Model Aeronautics in 1936. As airlines graduated in the 1930s from carrying the U.S. mail to transporting passengers, hobbyists continued to fly model aircraft in backyards, parks, and open fields.

In 1981, the FAA issued its Advisory Circular on Model Aircraft Operating Standards.[4] This one-page document encouraged model aircraft operators to select a site away from populated areas, not operate higher than 400 feet, notify airport operators if they were within three miles of an airport, give the right of way to "full-scale aircraft," and not operate in the presence of "spectators until the aircraft is successfully flight tested and proven airworthy." How a model aircraft operator was expected to prove the airworthiness of the aircraft in the absence of any airworthiness standards or rules was not explained.

Drones did not capture the general public's attention, however, until the use and profile of drones increased dramatically in the wars and conflicts in the Middle East beginning in the 1990s. In the public's eyes, the drones operated by the U.S. military were large, looking more like airline aircraft than model aircraft flown by hobbyists. The prominence of drones used for surveillance caught the attention of federal, state, and local governments.

In 2005, the FAA issued a memorandum on the use of drones by federal and state government agencies,[5] so-called public use operations. The FAA began issuing Certificates of Waiver or Authorization (COAs) to allow limited operations of unmanned aerial vehicles (UAVs) by federal and local law enforcement agencies and universities. The 2005 memorandum set out a process by which public agencies could obtain a COA by establishing airworthiness through FAA or Department of Defense certification, or by other means.

The FAA followed up in 2007 with a policy notice prohibiting commercial use of UAVs in the absence of airworthiness standards or rules.[6] The FAA noted that "about 50 companies, universities and government organizations are developing and producing some 155 unmanned aircraft designs," while U.S. military forces were using more than 700 drones in Iraq. At that point, the FAA had issued more than 50 COAs. Drone operators were directed to obtain an experimental category airworthiness certificate, but such a special airworthiness certificate permitted only research and development, marketing surveys, and crew training. Presaging what nine years later became Part 107, the FAA explained that it was examining the feasibility of creating a different category of small and slow drones defined by operations within the operator's line of sight.

ARE DRONES AIRCRAFT?

These early FAA pronouncements explicitly treated drones as aircraft. As aircraft, drones could be operated only if registered and with an airworthiness certificate. Not everyone agreed, however. The FAA was forced to litigate this question when it sought to impose a $10,000 civil penalty on a photographer. In 2011 the photographer had allegedly operated a drone with a camera on the University of Virginia campus in a careless and reckless manner, including under a tunnel, causing a person to take evasive action.

Raphael Pirker, the drone photographer in question, sought review by the National Transportation Safety Board. Although an administrative law judge agreed with Pirker's argument that the FAA had declined to regulate model aircraft, the board reversed, citing the plain language in both the 1981 advisory circular and the 2007 policy notice,[7] including the prohibition on commercial operations.[8] As discussed in the next section, Congress decided the question in the FAA Modernization and Reform Act of 2012.

The *Pirker* case illustrated the advances in camera technology: drones could take photographs and videos that couldn't be taken from the ground or even from a manned aircraft. Whereas a modeler would marvel at the dynamics of flight, the new drone operator would marvel at the wonders seen from a bird's eye view.

Photographers, journalists, and filmmakers all rushed to exploit drone technology. Drones were seen operating at weddings and other special events; and real estate agents supplemented virtual house tours made with hand-held cameras with virtual tours around and above the properties performed by drones.

While many other uses began to proliferate, it was drone photography that accelerated the pace of drone technology as well as the frequency of drone flights. But were these operations legal? Congress waded in for the first time.

HOW FAA REGULATIONS AND POLICIES AFFECT THE GROWTH AND BREADTH OF DRONE OPERATIONS

THE FAA MODERNIZATION AND REFORM ACT OF 2012

In the FAA Modernization and Reform Act of 2012, Congress required the FAA to develop a plan for safe integration of drones into the National Airspace System "as soon as practicable, but not later than September 30, 2015."[9] Some read this directive as requiring *integration* within a three-year period, but the more plausible reading is that Congress required only an integration *plan* by 2015. The FAA was also required to develop a roadmap, which it first issued in 2013.

Congress was aware that a few countries were ahead of the United States in establishing a regulatory framework for UAS and in authorizing UAS for commercial purposes. It was also aware that there were many unauthorized drone operations in the United States. Congress thus took two major steps to create a regulatory framework.

First, Congress directed the FAA to promulgate a rule for small UAS. The FAA had been developing this rule for a number of years and eventually published a final rule—adopting new Part 107—in June 2016, effective August 28, 2016. Second, in the interim period before a final rule was effective, Section 333 of the 2012 law allowed the FAA to grant exemptions, authorizing commercial operations of drones under 55 pounds and below 400 feet above ground level (AGL) without an airworthiness certificate. As discussed later, Section 333 was a success. By the time Part 107 took effect in August 2016, the FAA had issued over 5,500 exemptions.

Model aircraft operators perceived the 2012 law as a mixed bag. The statute defined "model aircraft" and in Section 336 prohibited the FAA from regulating "model aircraft" operated within certain parameters set forth in that section. While some model aircraft operators chafed at the restrictions in Section 336,[10] many commercial operators wondered why using the same make and model of drone, in the same airspace, for commercial purposes required a Section 333 exemption. Operators might wait many months to get FAA approval and, with that approval, be subject to many conditions and limitations, including the requirement that the remote pilot hold an airman certificate.

When the FAA issued an interim final rule in December 2015 requiring the registration of any drone over 55 pounds (including model aircraft as defined in the 2012 law), model aircraft operator John Taylor challenged it in court. The court of appeals in *Taylor v. Huerta,* 856 F.3d 1089 (D.C. Cir. 2017), held that Section 336 denied the FAA authority to regulate model aircraft and that requiring such aircraft to register was a statutorily prohibited regulation. This decision was controversial. The court of appeals did not apply the long-standing statutory requirement, codified in 49 U.S.C. § 44101(a),[11] that all civil aircraft be registered, in light of Section 336. Given that the FAA (and Congress in the very same 2012 statute) considered model aircraft to be "aircraft," the court of appeals essentially repealed the earlier statute's application to *model aircraft* without an unequivocal statement in the 2012 law to do so. The Supreme Court has held that repeals by implication are disfavored,[12] and this precedent was not discussed in the court of appeals opinion. Instead, the court of appeals found that the "FAA has not previously interpreted the general registration statute to apply to model aircraft."[13] Neither the 1981 advisory circular nor any other subsequent FAA policy statement expressly stated that model aircraft must be registered.

Alas, the United States did not seek review by the Supreme Court, and model aircraft operators were free to avoid registration as of May 2017. The UAS industry, joined by the airline industry, pushed Congress to overturn this decision. Congress did so in the National Defense Authorization Act of 2018,[14] reinstating the registration requirement retroactively. However, Congress did not repeal or revise Section 336.

SECTION 333 EXEMPTIONS

For commercial operators, Section 333 got off to a slow start. The first set of exemptions, for closed-set filming, were issued in September 2014, two and a half years after enactment of Section 333. The public nature of the petitions and the FAA's disposition of these petitions revealed the manifold commercial uses for small drones.

After granting exemptions for closed-set filming, the FAA began approving exemptions for precision agriculture. However, the FAA took much longer to grant exemptions to operate drones over 55 pounds for low-altitude agricultural spraying. Beyond agriculture, petitions sought to conduct wildlife management, cattle and livestock monitoring, and environmental remediation monitoring.

Operators began using drones for survey and inspection operations in a wide variety of industries and businesses. Drones were rapidly replacing manned aircraft, including helicopters, and taking the place of workers on ladders and other precarious settings. Drones could provide a more complete and accurate picture of the state of a bridge, tunnel, or road. Drones have undoubtedly saved many lives, spared workers career-ending injuries, and saved business hundreds of millions of dollars in the flare stack, telephone tower, railroad, and other industries.

Airport authorities, initially concerned about drone operations interfering with aircraft taking off from and landing at commercial service airports, warmed up to the potential benefits from using drones to conduct airport perimeter monitoring, inspecting runways and taxiways for foreign object debris, and even inspecting the aircraft fuselage. For any business, a drone performing perimeter security provides enhanced views over stationary closed-circuit television monitors.

Some Section 333 petitioners obtained exemptions to conduct search and rescue operations and otherwise assist first responders in the wake of a natural disaster or other catastrophe. Federal, state, and local governments were not required to obtain a Section 333 exemption to operate drones for these purposes, as public aircraft are generally not subject to federal aviation regulations. And as noted previously, the FAA had for years provided government agencies with COAs. Nongovernmental entities, both for-profit and nonprofit, arguably could be retained under contract with a public agency. However, many petitioners sought a Section 333 exemption in order to provide assistance to a number of jurisdictions without having to go through a lengthy procurement process with each such jurisdiction.

Insurance companies quickly realized the benefits of using drones to expedite insurance claims following a hurricane, flood, earthquake, or other disaster. Power and utility companies, as well as telecommunications providers, relied on drones to help restore infrastructure in communities.

In addition, local radio and television networks authorized by exemption to use drones for newsgathering expanded their operations to cover sporting events, such as skiing competitions, where drones could capture incredible images at much lower cost than stationary cameras.

Some drones can be used in aid of firefighting, but that use has not attracted the same media attention as the drones that were allegedly used recklessly to frustrate firefighters.[15] In 2016, Congress authorized a $20,000 civil penalty for a drone operator who interferes with wildfire suppression efforts (as well as any other first responder activity).[16] In the same law, Congress urged the FAA to grant exemptions on an emergency basis to permit drones to facilitate emergency response operations, including firefighting.[17]

Drones can assist firefighters in much the same way they assist police officers. A drone can detect problems that law enforcement and firefighters cannot see, whether because of distance or shielding. Drones can also carry light equipment to places where firetrucks and ladders cannot go. Park service and fire officials use drones in aid of vegetation management, which can be an effective fire prevention method.

PART 107 PUBLISHED TO AUTHORIZE COMMERCIAL OPERATIONS BY RULE

By the time the FAA published its proposed Part 107 rule in February 2015, the public health and safety benefits of small drone operations were evident. Still, the FAA's proposal to authorize commercial operations by rule rather than exemption or waiver was modest. At the behest of the UAS industry, the final rule published in June 2016 provided waiver authority beyond the authority commonly granted in Section 333 exemptions: to operate at night, beyond the visual line of sight of the remote pilot, and over people.[18]

For many commercial operators, Part 107 was all that was needed. Real estate showings, special events photography and videography, closed-set filming, and inspection of private property can all conceivably be conducted without a waiver. For some operators, the occasional flight of a drone over one or more persons at a closed event, such as a family gathering, can be conducted without a waiver, and outside the presence of any FAA or local law enforcement official.

However, many other commercial drone operators still require one or more waivers under Part 107. To enjoy myriad benefits at scale, they need permission for operations over people (OOP) and operations beyond the visual line of sight (BVLOS) of the remote pilot. Therefore, although Part 107 was put in place to obviate Section 333 exemptions, the rule's limits necessarily created a waiver process that could be regarded as Section 333 *redux*.

Section 333 exemptions and Part 107 waivers share one characteristic: each comes with a long list of conditions and limitations. Thus, for many, the conditions and limitations in a Part 107 waiver limit the scope of beneficial uses. In other respects, for drone operators as well as the public, the waiver process under Part 107 differs from the Section 333 process. Unlike the Section 333 process, where petitions and exemptions are publicly available, Part 107 waiver petitions are not available to the public, and waiver decisions typically do not provide much transparency on how the petitioner carried its burden of demonstrating an equivalent level of safety. That information is likely to be located in flight manuals, safety management systems, and other risk assessments that the petitioner regards as proprietary and thus provides to the FAA confidentially. Prompted by concerns expressed by the UAS industry,[19] the FAA has endeavored to provide guidance on seeking waivers under Part 107, on its website and at FAA UAS Symposia—in person and on webinars during the coronavirus pandemic.

The limitations of Part 107 waiver authority have come into sharp focus during the coronavirus pandemic. When the pandemic resulted in business shutdowns and stay-at-home orders in March 2020, the potential use of drones to deliver medical supplies and needed goods and foodstuffs became readily apparent. The Small UAV Coalition wrote to the Department of Transportation (DOT) secretary and FAA administrator on March 4, urging the use of waivers and exemptions to authorize drone delivery of medical equipment.[20] Medical equipment includes personal protective equipment, testing kits, and vaccines. Delivery of prescription drugs, food, and other supplies can also require a vehicle of some sort. Drone delivery promised not only expedited service but also contactless (touchless) service to avoid infection.

As of this writing, only a small number of drone operators have obtained waivers to deliver medical supplies BVLOS. This is likely because (a) obtaining a BVLOS waiver is time consuming; (b) operating—even with a waiver—is somewhat costly since visual observers are still required; and (c) many operations would entail operating over people, which the FAA might not approve. The FAA has developed an exemplary record of accomplishment in authorizing drone operations in the aftermath of hurricanes and other natural disasters, including the issuance of temporary flight restrictions to clear the airspace of noncooperative aircraft. However, the FAA does not regard drone operations responding to the pandemic by delivering medical equipment to have a sufficient nexus to the immediacy of disaster relief efforts, as contemplated in 14 C.F.R. 91.137.[21]

SECTION 44807 IS AVAILABLE WHERE PART 107 IS NOT

Part 107 is not available for operations using a drone, with payload, weighing more than 55 pounds, such as agricultural spraying; and there is no waiver provision in the rule for operations using heavier drones. At the time Part 107 took effect, Section 333 remained on the books as the regulatory avenue for these drones to use. It was superseded by U.S. Code Section 44807—Section 347 of the FAA Reauthorization Act of 2018.[22]

Package delivery, widely considered the use with the most business promise, does not make commercial sense unless it can be conducted BVLOS. But under Part 107, a BVLOS waiver may not be used to conduct package delivery for compensation or hire.[23] So, Section 44807, the successor to Section 333, is the gateway to package delivery for compensation or hire, requiring an air carrier operating certificate authorizing operations under Part 135.[24]

To date, only three companies have received a Section 44807 exemption from the airworthiness certification requirement and a Part 135 air carrier operating certificate: Wing, UPS Flight Forward, and most recently

Amazon Prime Air.[25] The FAA has noted in each case that its grant of exemptions is based in part on the progress each petitioner has made in obtaining type and airworthiness certification. While these developments are promising, the authority the FAA has granted each company is significantly circumscribed, both in terms of scope and in the conditions under which BVLOS operations are permitted. A charter air carrier operating transport category aircraft (weighing more than 12,500 pounds) and carrying passengers receives *nationwide authority* when it obtains a Part 135 air carrier operating certificate. By contrast, the UAS operators with a Part 135 certificate are confined to limited areas where the risk to persons or property on the ground is relatively low. With respect to BVLOS operations, UAS operators are required to have visual observers ensure the airspace is free of potentially conflicting air traffic, in the absence of FAA-approved detect-and-avoid (DAA) capability and perhaps BVLOS standards or rules. The set of operational and technical mitigations these petitioners have offered has not sufficed to enable BVLOS operations in urban and suburban settings without the aid of visual observers. In addition, these Section 44807 exemptions do not authorize operations over a congested area or assembly of people. Presumably such authority must await the issuance of type and airworthiness certificates to a company, or compliance with the recently published OOP final rule, which authorizes such operations in Categories 1, 2, and 4 of the final rule, subject to conditions and limitations.[26]

WHAT COULD SLOW THE GROWTH OF THE INDUSTRY? TECHNOLOGY IMPROVEMENTS AND REGULATIONS

The conditions and limitations the FAA initially included in Section 333 exemptions, and currently includes in Part 107 waivers and Section 44807 exemptions, reflect the current limitations of both UAS technology and UAS regulation. The four main technology imperatives—remote identification, DAA capability, aircraft certification, and UAS traffic management (UTM)—need to be solved to allow the UAS industry to scale and flourish, delivering social, economic, and public benefits to businesses and individuals.

REMOTE IDENTIFICATION

The FAA Extension, Safety, and Security Act of 2016 required the FAA to develop remote identification (ID) standards within one year and publish a rule within one year thereafter.[27] In May 2017, the FAA set up a UAS Identification and Tracking Aviation Rulemaking Committee, which issued its report at lightning speed in the fall of 2017. The task of developing standards was ceded to the American Society for Testing and Materials (ASTM) Special Committee F-38. The committee had substantially completed its document by the time the FAA published a proposed rule in December 2019. The UAS (and manned aviation) industry devoted enormous resources to the Aviation Rulemaking Committee and the ASTM efforts, as it had a strong incentive to do so. In December 2016, the FAA's proposed OOP rule was stalled. FAA Administrator Michael Huerta explained that further consultation with the national security and federal law enforcement community was necessary. It eventually became clear that no FAA UAS rule would be published until the remote ID rule was finalized.[28]

The FAA published its remote ID proposed rule on the final day of 2019.[29] More than 53,000 comments were filed in response, and the FAA faced several difficult issues. For example, should the rule require small UAVs to be equipped with broadcast *and* network capability, as proposed, or allow them to be equipped with either broadcast *or* network technology? DOT and the FAA pledged to publish a final rule by the end of 2020, and in fact did so, but surprised many in the UAS industry by requiring only broadcast technology and prohibiting network technology to comply with the requirements of the rule.[30]

Many companies seeking a Part 107 waiver or Section 44807 exemption include remote ID capability in their petition. The safety—and security and privacy—benefits from remote ID will come when compliance is near universal.

DETECT-AND-AVOID CAPABILITY

Foremost among the concepts that are critical to the safe integration of UAVs into the National Airspace System is DAA technology, which is akin to requiring that pilots see and avoid other aircraft. DAA technology will enable BVLOS operations without visual observers. This technology will prevent collisions with stationary and moving structures, including manned and unmanned aircraft. DAA technology may be ground based or onboard the drone. In uncontrolled airspace, not all manned aircraft are equipped with a transponder or other technology squawking a signal. Part 107 requires drone operators to give the right of way to manned aircraft, and thus the ability to detect such noncooperative aircraft is critical. For BVLOS operations of any considerable distance, ground-based DAA is not practical. The cost, similar to the cost of placing visual observers at three-mile intervals throughout a geographic area, would be prohibitive. Onboard DAA is the only option that will allow the industry to scale in both distance and frequency.

Several companies have developed onboard DAA technology, but the FAA has not yet approved use of DAA technology in BVLOS operations in a waiver or exemption. ASTM Special Committee F-38 has published a DAA performance standard.[31]

AIRCRAFT CERTIFICATION

In 2012 and again in 2018, Congress authorized the FAA to permit commercial UAS operations without the drone having an airworthiness certificate. Nevertheless, the FAA is committed to reducing the need for exemptions and waivers by adopting type and airworthiness standards for small UAVs and has been working with a number of UAS companies.

The FAA announced that it would use its existing authority in 14 C.F.R. 21.17(b) to designate some UAVs as a "special class" and certify them through special conditions.[32] This regulation recognizes that some models of aircraft have unique, novel, or unusual features that do not neatly fit within the parameter of Parts 23 (small aircraft) or 25 (transport category aircraft), and thus the FAA develops a set of special conditions to ensure airworthiness. While special conditions are aircraft-type-specific, the FAA is also developing general guidance. In late November 2020, the FAA proposed a set of airworthiness criteria for 10 UAS models.[33] These proposed criteria set out performance standards and require durability and reliability testing. Although not included in the proposed criteria, the FAA has developed an hours-and-population-density matrix as a compliance measure to demonstrate a sufficient degree of durability and reliability.

Notwithstanding this progress on UAS certification, the FAA must continue to use its Section 44807 exemption and Part 107 waiver authority to authorize BVLOS operations in the interim period. The FAA Reauthorization Act of 2018 sunsets the secretary's authority under Section 44807 on September 30, 2023. This sunset is prospective. Any authority granted before that date continues to exist for the duration set in a waiver or exemption. Mindful of the protracted duration of standards development and rulemaking, Congress may need to extend or revise Section 44807 before its expiration.

UNMANNED TRAFFIC MANAGEMENT

For the small UAS industry to thrive at scale, there must be a system to ensure safety and efficiency of multiple drones operating in the same low-altitude, uncontrolled airspace and to ensure the safety of drones and manned aircraft moving between controlled and uncontrolled airspace. This system will not be operated by air traffic controllers, who by definition do not separate or "de-conflict" aircraft in uncontrolled airspace. Rather, based on a concept developed by NASA with the participation of the FAA and the UAS industry, a cloud-based traffic management system will perform the services typically performed by controllers for manned aircraft.

Although the unmanned traffic management (UTM) concept has been in development for more than five years, the FAA is still in the early stages of development and implementation. The UAS industry, as with other technological challenges to UAS integration, has devoted many personnel and resources to developing a UTM system. Following NASA's research and development efforts and prodded by Congress, the FAA embarked on a UTM system pilot program, which remains pending.[34] The FAA has developed a UTM concept of operations document and published version 2.0 in 2020.[35] The FAA expects to publish version 3.0 in 2021.[36]

STATE AND LOCAL LAWS AND THE FUTURE OF DRONE POLICY

As previously stated, technology challenges and regulatory delays are likely to hamper the growth and profitability of UAS operations for several more years. It is axiomatic—and acknowledged by DOT and the

FAA—that regulations do not move at the pace of technology. In addition, in a few important areas noted earlier, technology needs further maturation.

Perhaps a greater and longer-term concern relates to state and local laws, which include both statutes and ordinances and state tort law. An in-depth analysis of state and local laws is considered in other chapters. This next section argues that changes to tort law could significantly limit low-altitude drone operations and prevent or hinder a variety of beneficial use cases.

Some believe that resistance to drones will come primarily from citizens voicing privacy, noise, and nuisance concerns. Certainly, state and local restrictions on drone operations reflect these concerns. However, drones will not become ubiquitous overnight; what may seem today to be a novel way of delivering goods or conducting surveys and inspections will eventually be so commonplace that many operations will go unnoticed. Public acceptance will increase concomitantly with greater exposure to the many beneficial uses discussed in this chapter. Hence, once the regulatory framework includes standards and rules for the key technological advances, such as onboard DAA equipment and UTM, drones will be woven into the fabric of civil society.

STATE AND LOCAL LAWS AND FEDERAL PREEMPTION

Many state and local governments have already enacted laws that restrict or regulate UAS operations and operators, with most of the focus to date on the use of UAVs by law enforcement agencies.[37] (State and federal law enforcement use of UAVs implicates the Fourth Amendment, which is discussed in Chapter Five.) Some state and local governments adopted drone laws purportedly as a placeholder until the FAA adopted a regulatory framework, but it is doubtful that any state or locality has repealed its law since the FAA promulgated Part 107. Local efforts to prohibit or restrict drones have focused on personal privacy and nuisance concerns when small drone operations are authorized to fly in residential and commercial districts and operations proliferate.[38]

The FAA recognizes that state and local governments are vested with police powers under state constitutions, which include zoning and land use, condemnation, and law enforcement.[39] At the same time, the FAA has asserted jurisdiction from the ground up, given that small drones may take off and land at nearly any outside location and do not need a typical airfield.[40] Moreover, FAA authority over aircraft (registration, certification, inspection, maintenance), airmen (training and qualification of pilots), air carriers (certification, regulation, inspection, and enforcement), and airspace (flight altitudes, flight paths) is plenary; no state or local government has regulatory authority over these subjects in interstate commerce.[41]

Would these state and local laws survive a challenge under the U.S. Constitution's supremacy clause? The jurisprudence on preemption of state and local laws under the Federal Aviation Act that has developed over decades is set out in the appendix to the FAA's 2015 Fact Sheet.[42] With respect to drones, there are only two reported decisions to date: one decision found preemption, the other found no preemption.

In *Singer v. Newton,* 284 F. Supp. 3d (D. Mass. 2017), the City of Newton, Massachusetts, adopted an ordinance that, in pertinent part, requires all drones to register with the city, prohibits BVLOS operations, and prohibits operations over city property without city permit and operations over private property below 400 feet AGL without consent. The district court held that the Federal Aviation Act did not preempt the field of drone regulation and struck down Newton's ordinance under conflict preemption. The registration requirement conflicts with FAA's registration rule. The prohibition on operating over city property has no altitude limit, so it conflicts with FAA's regulation of navigable airspace. The prohibition on BVLOS operations is also broader than Part 107, as it does not allow for any waivers. The court added that combining the bans over public and private property conflicts with Part 107 as to authorized flights.[43]

In *National Press Photographers Association v. McCraw,* 2020 WL 7029156 (W. D. Tex. Nov. 30, 2020), a newsgathering association and two reporters challenged two provisions of the Texas Government Code. One provision makes it unlawful to capture an image of an individual or privately owned real property with the intent to conduct surveillance on the individual or property. This provision exempts drones operated for educational purposes but not newsgathering. The so-called no-fly provision makes it unlawful to fly a drone less than 400 feet AGL over a correctional or detention facility, critical infrastructure facility, or sports venue. This provision exempts drones operated for "commercial purposes," a term that is not defined. On a motion to dismiss, the district court found that the plaintiffs made a sufficient showing that these provisions violate the First Amendment because they are content based, vague (i.e., the terms "surveillance" and "commercial purposes" are not defined in the Code), and overbroad. However, the district court granted the motion to dismiss plaintiffs' preemption claims, holding that the Federal Aviation Act does not occupy the field of drone regulation, and that the no-fly provision does not impermissibly conflict with FAA drone regulations.

Congressional efforts to include a preemption provision in federal law, whether to confirm or limit preemption, have not been successful to date. Time will tell whether the implied preemption precedents developed over the decades will apply with equal force with respect to drone operations and operators. The express preemption of state and local economic regulation (rate, route, and service) of air carriers, which was included in the Airline Deregulation Act of 1978,[44] should protect UAS air carriers from state and local economic regulation.

The drone industry should be able to thrive under current tort law, but that could change if efforts to change tort law succeed. The tort provisions that are most likely to apply to drone operations are trespass, nuisance, and invasion of privacy.

TRESPASS

Trespass is a common law tort arising from a property interest and may be codified in statute. Section 158 of the *Restatement of the Law, Second, Torts*[45] provides that trespass is a dignitary tort: harm is presumed and need not be proved. In the comment to section 158, several examples are given of trespasses including flying in an airplane over another's house close to the roof, propelling or placing a thing in the airspace above the property, and firing projectiles or flying an advertising kite or balloon through the airspace above the property.

In *United States v. Causby,* 328 U.S. 256 (1946), the Supreme Court held that the U.S. military took easement from Causby by operating aircraft over his chicken farm so low and so frequently that it deprived him of the use of his land. In that Court ruling, Justice William Douglas said the common law maxim *ad coelum*—that a landowner owned everything above and below the land—"has no place in the modern world."[46] But neither Congress, which established the navigable airspace and the public highway, nor the FAA took the easement; it was the airport owner, as was made clear later in *Griggs v. County of Allegheny,* 369 U.S. 84 (1962). Justice Douglas continued:

> Yet it is obvious that if the landowner is to have full enjoyment of the land, he must have exclusive control of the immediate reaches of the enveloping atmosphere. . . . The landowner owns at least as much of the space above the ground as he can occupy or use in connection with the land.[47]

As many observers have commented, Justice Douglas used some words loosely, without elaboration. His statement that a landowner must have "exclusive control" has been read to support obtaining an injunction to prevent drone overflight. However, a takings clause case is for money damages, not injunctive relief. Had Causby sought to block U.S. military flights during World War II, it is highly unlikely he would have prevailed.

What are the immediate reaches? Any attempt to draw a line in the sky is quixotic; an inverse condemnation case is very fact specific. After *United States v. Causby,* Congress revised the definition of "navigable airspace" to mean "airspace above the minimum altitudes of flight prescribed by [FAA] regulations . . . *including airspace needed to ensure safety in the takeoff and landing of aircraft.*"[48] However, the Supreme Court held in *Griggs* that a taking may occur in navigable airspace, confirming the protean notion of "immediate reaches."[49]

In 1965, the American Law Institute (ALI) recognized an aerial trespass tort in Section 159 of its *Restatement* and adapted this provision from the holding and much of the reasoning in *United States v. Causby.* Section 159 states that a trespass may be committed above the surface of the earth and provides, "Flight by aircraft in the air space above the land of another is a trespass if, but only if, (a) it enters into the immediate reaches of the air space next to the land, and (b) it interferes substantially with the other's use and enjoyment of the land."[50]

"Immediate reaches" was not defined in the *Restatement* or subsequently. In a comment on Section 159, the *Restatement* conflated the concepts of "immediate reaches" and "substantial interference." The comment states, "In the ordinary case, flight at 500 feet or more above the surface is not within the 'immediate reaches,' while flight within 50 feet, *which interferes with actual use*, clearly is, and flight within 150 feet, *which also so interferes*, may present a question of fact."[51]

While the drafters of Section 159 did not contemplate small drone operations, the aerial trespass tort is flexible enough to provide a remedy for drone operations that substantially interfere with a landowner's use and enjoyment of the land, while not providing a remedy solely because of an evanescent overflight. Others, however, believe that tort law needs to be revised to address low-altitude drone operations.

In 2015, the Uniform Law Commission (ULC) established a Drone Tort Committee to examine not only trespass but also other torts with respect to drone operations. The ULC is a group of judges, professors, and lawyers who agree to address a developing area of law, where the precedents may be few or conflicting, with the objective of proposing a model bill to be adopted by the states. The Drone Tort Committee invited UAS industry observers to participate in the two-year development of a Uniform Tort Law Relating to Drones Act. The initial draft created a 200-feet-AGL line in the sky, below which a drone operation would commit a per se trespass, without any showing of interference, much less "substantial" interference, per Section 159 of the *Restatement*. Opposition from the UAS industry, supported by a letter from DOT and the FAA, resulted in a decision by the ULC Plenary in the summer of 2018 to reconsider the per se trespass concept.

By May 2019, the Drone Tort Committee had developed an alternative draft, which abandoned the line-in-the-sky concept in favor of a set of nonexhaustive factors for a court to consider in determining whether a drone flight (or series of flights) substantially interfered with the landowner's use or enjoyment of the land.[52] The draft also removed the "immediate reaches" concept as unworkable and unnecessary in light of the factors.

The factors "potentially relevant" to the question of "substantial interference" are as follows:

1. nature of use and enjoyment of the property;
2. the operator's purpose in operating over the property;
3. the altitude of the aircraft;
4. the amount of time the aircraft is over the property;
5. the frequency with which drones have operated over the property during the relevant period;
6. the type of drone and nature of its operation over the property;
7. whether the drone directly caused physical or emotional injury to persons/damage to real or personal property;
8. whether the drone directly caused economic damage;
9. the time of day of drone operation over the property;
10. whether an individual on the property saw or heard the drone while it was over the property;
11. whether and to what extent to which the operation exceeded any consent of the land possessor;
12. whether the drone harassed persons, livestock, or wildlife on the property, regardless of purpose of operation; and
13. any other relevant factor.[53]

The draft also noted that "[r]epeated or continual operation of drones over a land possessor's property does not create a prescriptive right in the airspace."[54] It included a rebuttable presumption that a drone operation does not substantially interfere with the use and enjoyment of property if the drone is operated for

- law enforcement in conformance with the Fourth Amendment, including any warrant or court order;
- purposes protected by the First Amendment; or

- purposes of protecting public safety by authorized personnel in emergency situations.[55]

The UAS industry supported this revised provision, but it was not adopted by the ULC Plenary meeting in the summer of 2019. Some members of the commission expressed concern that the Drone Tort Committee's revised proposal did not sufficiently respect private property interests.

Months later, the UAS industry learned that the ALI, as part of its work on the *Restatement of the Law, Fourth, Property*, had developed a "trespass by overflight" provision. Like the ULC, the ALI comprises judges, professors, and attorneys. Unlike the ULC, whose objective is to create a model statute for states to adopt, the ALI purports to be simply "restating" the law to assist courts and practitioners. However, as the "trespass by overflight" provision illustrates, the ALI has on occasion strayed into shaping new law under the guise of not doing so.

The draft ALI provision would not create a line in the sky but would give a landowner airspace rights above the airspace that the landowner is now using (e.g., for a building) to seek money damages and injunctive relief. If this "trespass by overflight" is adopted, courts may be inclined to rely on this provision and ignore the aerial trespass provision in the *Restatement of the Law, Second, Torts.* The draft provision was removed from consideration at the ALI 2021 annual meeting held virtually in May and June, and may be revised before it is submitted for adoption in 2022.

NUISANCE

Existing nuisance law should not hinder the UAS industry. A private nuisance, as set out in Sections 821D and 821F of the *Restatement* (vol. 4) is a nontrespassory invasion of another's interest in private use and enjoyment of land. That is to say, there is no trespass, but there still is interference. In a comment to Section 821D, the term "interest in use and enjoyment" of land "comprehends the pleasure, comfort and enjoyment that a person normally derives from the occupancy of land."[56] There is no liability without significant harm, and thus the interference must be more than an inconvenience or petty annoyance. The comment notes that interference must be intentional or unreasonable. The private nuisance tort, as applied to drone operations, is quite similar to the aerial trespass provision. It applies to drone operations that do not amount to a trespass.

INVASION OF PRIVACY

The UAS industry recognizes that drone operations could invade a person's reasonable expectation of privacy but does not believe the existing tort law will inhibit low-altitude drone flights at scale.[57]

Invasion of privacy, also referred to as intrusion upon seclusion, is an intentional tort. Section 652A of the *Restatement* (vol. 3), which sets out general principles, identifies four types of invasions. "Unreasonable intrusion upon seclusion of another" applies in the drone setting. "One who intentionally intrudes upon the solitude or seclusion of another or his private affairs or concerns, is subject to liability to the other for invasion of his privacy, if the intrusion would be highly offensive to a reasonable person."[58] The comment to Section 652B notes that the interference must be substantial, thus making this tort consistent with aerial trespass and private nuisance.[59]

State and local governments may seek to redefine an invasion of privacy by a drone in a state statute or local ordinance and provide a private right of action to supplement the common law of torts in that jurisdiction. Depending on the specifics of such invasion of privacy statute, a drone operator may be able to challenge the law as impliedly preempted should it conflict with the Federal Aviation Act and rules promulgated thereunder, such as Part 107. A state or local government's objective of protecting citizens' privacy does not immunize the law from federal preemption.

CONCLUSION

The theme for the FAA UAS Symposium held virtually in the summer of 2020 was "Drones Are Here for Good." That theme sums up this chapter. Drones will increase in number to accommodate the many uses envisioned by businesses and demanded by the public. The FAA will complete a regulatory framework in which drones may safely and reliably operate autonomously, over people, and BVLOS. Every step forward taken by the UAS industry and the FAA will open up greater opportunities for drones to save lives, reduce the risk of injuries to workers, and make work and pleasure activities more efficient while reducing logistics' carbon footprint. While many challenges remain, the future of drones is bright. Drones are here for good.

CHAPTER THREE

REFRAMING DRONE POLICY TO EMBRACE INNOVATION IN AMERICA

JAMES CZERNIAWSKI

While militaries have a history of working with drones that dates back more than a hundred years, commercial drone use is relatively new.[1] The Federal Aviation Administration (FAA) issued the first commercial drone permit only in 2006.[2]

From photography to environmental applications to weapons of war, drones have made their way into numerous aspects of our lives.[3] However, a whole litany of federal, state, and local laws makes navigating the drone industry a challenge.[4] Operating in such a complicated legal environment will leave both companies and consumers worse off in the long run.

Yet despite all the current and perceived future benefits of drones, lawmakers have not fully embraced the technology. And so, the drone industry faces many regulatory hurdles. Like other emerging technologies before it, the drone industry finds itself in regulators' crosshairs.

Anxiety about new technologies is not new. In the 1860s, the British government passed a series of Locomotive Acts out of fear of the perceived risks of early automobiles. The Locomotive Act of 1865 was the most egregious. It included a "red flag" provision, which required each vehicle to be manned by a crew of three, one of whom needed to walk in front of the car with a red flag during the daytime or a lamp at night. Additionally, the law created speed limits of 4 miles per hour in the countryside and 2 miles per hour in towns. The government fined violators £10 for noncompliance.[5] That disastrous policy decision set back the United Kingdom decades and ultimately stunted growth in the industry.[6] The concern today is that drone companies have never been able to take off owing to similarly oppressive regulations.

Drone technology can increase prosperity in the United States and worldwide, but it needs sound policy and institutions to do so. Martec's Law states that technology grows at an exponential rate, whereas the government's ability to change with the technology is more logarithmic.[7] Allowing this rigid and lethargic law of regulation to govern drone technology prevents people from enjoying the technology's benefits today. Unsurprisingly, companies believe that regulatory change is a primary driver of their ability to offer their services.[8] The reality is that drones are increasingly popular and will play a growing role in our everyday lives, both commercially and on an individual level. Fearmongering over the technology is not the answer. If we follow a least restrictive means framework, pragmatic regulations can help unleash the full potential of drone technology.

In this chapter, I begin by looking at the benefits drone technology has provided across numerous industries. Then I examine some of the core drone technology policy controversies in the United States. Finally, I discuss potential policy solutions to improve U.S. drone policy to bring it more in line with the rest of the world.

USE CASES OF DRONE TECHNOLOGY

AGRICULTURE

PricewaterhouseCoopers projected in 2016 that drone-driven solutions in the agriculture industry would be worth approximately $32 billion.[9] Drones can have an impact on a variety of critical areas for the agricultural sector.

For example, drones can monitor soil and conduct field analysis. By generating exact three-dimensional (3D) maps, drones can help farmers plan seeding patterns.[10] Drones can be beneficial in the planting phase of crops. Drones manufactured by startup companies can achieve an uptake rate of 75 percent and reduce planting costs by 85 percent. Drones can deploy pods filled with seeds and nutrients directly into the soil.[11] Widespread adoption would be a significant step toward reducing the price of food.

Crop monitoring is one of the most daunting challenges farmers face. Rather than relying on satellite imagery, which can be imprecise and involve significant time delays, farmers can use drones, which offer real-time analysis.[12] Drones equipped with specific sensors can more efficiently monitor crops by identifying dry areas and calculating the vegetation index, examining the plant health within a geographic area.[13]

Additionally, drones can monitor the overarching health of the crop and recognize potentially sick or disease-ridden plants. By utilizing the sensors to monitor the plants' health, farmers can rapidly respond using precise remedies. Proactively monitoring crops could play a significant role in saving crops sooner, leading to less waste.[14]

Drones are actively helping farmers around the world today. For example, in a soybean field in Brazil, farmers used drones to take accurate, high-quality images of their land. Running the images through farming software, they were alerted to areas of their fields infested with weeds. The farmers generated herbicide application maps based on the drone information and saved over half of the herbicides used by other farmers during the 2018–2019 farming season in Brazil.[15]

The future presents many challenges for farmers. Growing worldwide populations that need more food, coupled with constraints on the water supply, further reinforce the need for farmers to seek out new ideas.[16] Drones can empower farmers with innovative solutions to help tackle problems in their industry. Agriculture drones can help farmers more efficiently manage crops and decrease input costs, leading to lower food prices for consumers.

ARCHITECTURE AND CONSTRUCTION

The architecture and construction industries could benefit from incorporating drones into their operations, and demand in both sectors is growing. Utilizing drones in a variety of roles can help keep projects on time and budget.

Drones can help architects early in the design and construction process. Unlike traditional mapping methods, drones can scan a large geographic area, generating high-resolution 3D images of a site and the surrounding area. The information provided by drones can be extremely useful in determining how project managers will develop the land.[17]

In particular, the data drones generate can help site managers determine how many and which materials are needed. Thus drones create opportunities to cut costs by giving a clearer understanding of what a particular construction project requires.[18]

Drones also help hold the construction industry accountable. Because drones can monitor sites efficiently, they can serve in a preventative role, allowing site managers to correct mistakes earlier in the construction process. Additionally, drones can facilitate a more efficient labor flow within a worksite, mitigating many potential safety hazards in the process.[19]

Another fascinating use of drones is in the construction process itself. For example, in 2015, researchers in Germany used drones to autonomously construct a "lightweight tensile bridge" capable of supporting a human's weight. Using material with a low weight-to-strength ratio, drones built a bridge 7.5 meters long.[20] In 2012, Swiss architecture firm Gramazio Kohler and roboticist Raffaello D'Andrea used drones to stack thousands of polystyrene blocks.[21]

Engineers and architects have only just begun using drones, but so far, the results are promising. Increasing efficiency and productivity, decreasing costs, and potentially increasing worker safety are all noteworthy. Each of these aspects could contribute to reducing housing costs across the United States—an increasingly prevalent problem in many of America's major cities.

DELIVERY

Another area where drones can make an impact is delivery services. Delivery is a significant part of the global economy, with over 5.4 billion parcels delivered in just the United States in 2019.[22] That same year, the international parcel delivery market was worth approximately $430 billion, up from $380 billion the year before.[23] The coronavirus pandemic shined a light on the benefits of drones used in this capacity, laying the groundwork for the continued growth of delivery drones in the years to come.[24] With global online retail sales expected to continue to grow in both developed and emerging economies, the sector's workload will likely increase. Many companies have expressed a desire to utilize drones for delivery, which could have wide-ranging implications for consumers and entrepreneurs.

For example, in 2013, Amazon chief executive Jeff Bezos announced in a CBS *60 Minutes* interview Amazon's intention to use drones for parcel delivery. Always looking for ways to change the parcel delivery industry, Bezos envisioned drones for delivering goods that weighed less than five pounds in an estimated 30 minutes. He projected that 86 percent of Amazon products would qualify for drone delivery under that criterion.[25] One of the primary reasons the program struggled to take off since then is that Amazon has gotten bogged down, trying to work with regulators at the FAA to take this idea and turn it into a reality.[26] Ultimately, the company took its drone operations overseas to England, where the regulations allowed such flights to occur. In December 2016, the company successfully used a drone to deliver its first package in Cambridge, taking 13 minutes to carry the order from the fulfillment warehouse.[27]

Another early mover in the drone industry was Alphabet, the parent company of Google. In 2012, Alphabet created the subsidiary Wing as part of its X Development initiative. The drone developed by Wing was very impressive, with the ability to hover and winch packages to and from the ground with a retractable tether.[28]

Wing has had a promising start. The company has delivered more than 5,000 packages in Australia, Finland, and the United States. From donuts and cheese to FedEx packages, Wing has brought a wide variety of goods to consumers' doors. But despite proving that drones are a viable delivery option, Wing has run into issues similar to Amazon's. The company can complete deliveries but is severely limited in where it is allowed to do so. In Australia, the service is limited to two cities; in Finland, only Helsinki residents have access; and in the United States, the town of Christiansburg, Virginia, was the first town to allow drone delivery of goods.[29]

Drones can also address infrastructure issues. In cities, one of the most common problems is congestion, while for more rural areas, the lack of infrastructure is equally problematic. Drones present a solution to both issues. Drones can alleviate congestion by reducing the need for ground-based delivery vehicles.[30] By extension, local governments benefit because fewer ground-based delivery vehicles on the road mean less wear and tear on the streets, making them easier to maintain.

Many of the emerging drone delivery services aim to deliver goods in 30 minutes or less, an admittedly ambitious goal. However, if companies can consistently deliver consumer goods promptly, that is a valuable service to have at their disposal. Customers stand to benefit immensely, getting the goods they want, when they want them—even time-sensitive goods such as medicine.[31]

Drones can also assist with heavier cargo shipping. Cargo drones are one of the newest developments in the drone industry. They are capable of lifting hundreds of pounds and traveling hundreds of miles. As an added benefit, they are also more economically friendly than the traditional methods of shipping.[32]

Numerous companies both large and small are in a race to develop these long-range cargo drones. The drones they are developing range in size, shape, and speed. For example, Boeing's cargo drone weighs more than 700 pounds and can carry up to 500 pounds; Sabrewing's cargo drone will reach speeds of just over 200 miles per hour.[33]

Not surprisingly, these companies are pushing hard to develop these kinds of drones. Among other opportunities, they see the ability to partner with traditional carriers as highly lucrative. Morgan Stanley estimates that the drone cargo industry could be worth as much as $1.5 trillion by 2040.[34]

Many of the drone industry's examples in delivery services display the range and variety of drones' abilities to significantly improve American lives. Decreasing shipping costs, a more environmentally friendly vehicle, and the ability to get consumers products faster than ever before all point to a promising future.

DEFENSE/SECURITY

It is crucial to understand the role drones have played in the military, law enforcement, and security. As already mentioned, the military has been using drones for well over 100 years. The U.S. military has indicated its seriousness by investing $2.2 billion in research and development of drones in 2020 alone, increasing to $2.7 billion by 2029. On top of that, the U.S. military expects to spend as much as $3.3 billion to procure drones by 2029.[35]

One of the primary military uses of drones is surveillance, which provides troops on the ground with potentially life-saving intelligence. The newest drone in this vein is called the Ultra LEAP (for Long Endurance Aircraft Platform), which has logged more than 18,000 combat hours.[36] The military uses this drone and others for reconnaissance. Such drones can stay in the air for an extended period to provide live footage and photos. That critical information can help troops prepare before arriving on the scene. The information drones provide also plays a part in a process known as battle damage assessment, a type of reconnaissance that enables the military to estimate the damage done by a military force in a previous engagement. This assessment can potentially influence critical military decisions.[37]

The most cynical use of drones by the military is as a weapon: drones can carry out attacks on targets that would otherwise be conducted by manned aircraft. Using drones in this way has become an increasingly attractive option for the military. President George W. Bush authorized approximately 54 drone strikes throughout his presidency.[38] During President Barack Obama's two terms, he normalized the practice, authorizing more than 540 drone strikes, 10 times more than his predecessor.[39] These weapons have certainly hit their mark, but they have also killed innocent people. Micah Zenko, author of the book *Clear and Present Safety*, estimates 324 civilians died in the drone strikes President Obama authorized.[40]

Law enforcement agencies' use of drones varies slightly from their military counterparts. For example, one way that law enforcement uses drones is to map cities. The drones analyze the city from above and look for the densely populated areas. Such use of drones could prove helpful when law enforcement needs to allocate resources in a natural disaster to know where to respond first.[41]

Law enforcement uses drones at crime scenes to analyze the scene and provide a 3D map and images in mere minutes. Investigators also benefit from capturing footage from previously unattainable angles without spending an excessive amount of money on a helicopter to fly above the scene. Rather than spending as much as $600 an hour for a helicopter to navigate the city, police departments can spend a fraction of that on acquiring drones to perform similar tasks.[42]

Unfortunately, some law enforcement agencies across the country have used drones for less laudable reasons. Take what has occurred during the coronavirus pandemic. In New York, for example, law enforcement used drones to spot and fine residents to enforce lockdown orders.[43] California and New Jersey officials have also used drones to monitor and enforce social distancing during lockdowns. New Jersey police officers used the technology to spot noncompliant citizens and fine violators up to $1,000.

Actions like these are intended "to help combat people not following social distancing," according to one police department. While that may seem like a harsh invasion of privacy, a New Jersey police chief assures us, it's worth it if it "saves one life."[44] Yet law enforcement, in this author's opinion, should not be in the business of leveraging technologies in ways that fundamentally violate reasonable expectations of privacy in the Fourth Amendment of the Constitution.

Research and development of drone technology for military and law enforcement uses will continue to be common. It is critical to remember the unique issues that come with these uses contrasted against the industries mentioned earlier. Understanding the differences between use cases can lead to more tailored policy decisions.

EMERGENCY SERVICES

Drones can play a vital role in the emergency services sector. One area where drones can be beneficial is in the event of a fire. With drones equipped with thermal imagery sensors, firefighters can more accurately place fire retardant where it is most needed.[45]

Drones can be helpful during other crises as well, such as when a natural disaster occurs. Drones have been used to conduct search and rescue missions during disaster response efforts. Da-Jiang Innovations, known as DJI, is a leading Chinese drone manufacturer that claims its drones are responsible for "saving the lives of at least 279 people."[46]

On September 16, 2017, Hurricane Maria hit the island of Puerto Rico as a Category 4 storm, with winds of up to 155 miles per hour.[47] The storm laid waste to the island, destroying most of its power lines and crippling its infrastructure.[48] A study conducted by George Washington University estimated that 2,975 Puerto Ricans died.[49] When natural disasters occur, an essential element in coordinating the response and recovery efforts is the establishment of cells on wheels (COWs) to provide critical communications services when ordinary cell service is down. However, due to the severity of the hurricane, the usual methods for setting up COWs were not feasible. At one point, more than 90 percent of cell towers in Puerto Rico were down. The storm also made many roads impassable. The lack of communications infrastructure was significantly slowing down recovery efforts.[50]

Recognizing the dire nature of the situation, AT&T, a major telecommunications company, took action. The FAA granted a special permit to utilize *flying* COWs—cells on wings. These drones were able to fly over Puerto Rico and broadcast voice, data, and Internet service. Each drone could cover approximately 40 square miles. The primary reason AT&T needed FAA approval was that the drones weighed more than 55 pounds—the limit for small drones.[51] Many such restrictions remain common in the industry.

Drones have played a critical role in delivering medical supplies during the coronavirus pandemic. In Ireland, a drone delivered insulin to a patient on a remote island, which had never happened before. The drone also brought back a blood sample for monitoring glucose.[52] This proved an effective way of transporting critical medical goods while simultaneously maintaining social distancing measures to help curb the virus's spread. In another pandemic-related example, concerns were raised early on about potential crop contamination with COVID-19. Drones were used to spray disinfectant on crops, thus protecting consumers while not needlessly exposing agricultural workers to the virus.[53]

Drones can play a critical role in a wide range of disaster response and recovery efforts. From search and rescue to communications infrastructure to pandemic mitigation, drones have been used with great success.

MEDIA AND ENTERTAINMENT

Media companies have found and are contemplating various exciting ways to use drone technology to improve the product they provide to consumers across numerous fields. The multifaceted capabilities of drones enable them to help multiple professionals in the media industry.

Take the movie industry, for example. The FAA formally allowed the sector to utilize the technology in filming in 2014, and it promptly capitalized on the opportunity. Cinematographers realized early on that drones provide an excellent way of capturing intense action sequences. A great example is the opening motorcycle chase scene in the hit James Bond movie *Skyfall*. Another is *The Wolf of Wall Street*; Martin Scorsese used drones for an overhead shot of a party scene, effectively conveying the party's energy.[54]

Drones have also influenced the ways viewers experience films, providing different perspectives. For example, in *Jurassic World*, drones equipped with cameras simulated the pterosaurs' point of view as they were attacking humans. In an episode of the television show *Black Mirror*, filmmakers used drones to highlight the privacy concerns drones could present: An individual blackmails the protagonists by filming them with a drone to capture private moments of their personal lives. Throughout the episode, viewers see traditional camera shots, as well as shots from the drone's perspective, casting light on people's privacy concerns about the technology.[55]

As drones overcome certain technological hurdles, drones will become increasingly prevalent in the film industry. They provide similar quality shots as could be captured from a helicopter or plane, at a fraction of the cost. Estimates for filming from a helicopter range from $20,000 to $40,000, while the cost of using a drone is $4,500 to $13,000. The cost savings could do many favors for an industry always looking for new and innovative ways to adapt their production processes.[56]

PROBLEMS WITH DRONE POLICY IN THE UNITED STATES

Drones have the potential to do good across various sectors, with consumers and producers alike benefiting from the technology's numerous capabilities. Yet in the United States, achieving widespread adoption of the technology has been extraordinarily challenging. A few glaring issues in U.S. drone policy are hampering the industry from unlocking its true potential.

Congress created the Federal Aviation Administration in 1958 to "regulate aviation safety, the efficiency of the navigable airspace, and air traffic control, among other things" for aircraft.[57] However, when it came to the potential regulation of hobby model planes, the FAA declined to implement regulations.

The rising popularity of drones, coupled with mounting pressure to take action on the technology, finally led the FAA to regulate these unmanned aerial vehicles (UAVs). In 2007, the FAA issued a statement clarifying its policy, interpreting drones as falling under the same legal definition as aircraft. In doing so, the FAA essentially subjected the entirety of the drone industry to its regulatory regime.[58] Suddenly, drone operators found themselves crushed by the weight of mandatory FAA regulations. From pilot qualification requirements (requiring drone operators to pass a test for flying a manned aircraft) to operation requirements (such as flying at low altitude and remaining in the operator's line of sight), drones were in a radically different position than they had been in the past.[59]

While consumer-oriented drones had their battles with the FAA, commercial drones did not fare much better. One of the monumental regulatory hurdles commercial drone companies face is Part 107 of the FAA regulations, which limits the weight of drones to 55 pounds or less.[60] The FAA coupled the weight requirement with both line of sight requirements and mandatory pilot certification requirements.[61] The regulations also banned drones from flying at night, flying more than 400 feet above ground level, traveling at speeds over 100 miles per hour, or operating around any large groups of people.[62]

In following this regulatory regime, the FAA is essentially cutting off the emerging industry at its knees, thwarting many practical applications of the technology.

Another issue with regulating drones and related technology with vertical takeoff and landing capability is the management of air traffic conditions. Regulators wishing to follow the system currently employed for managing airplanes may find that such a centralized process leads them to trying to fit a square peg in a round hole.

Even Congress recognized the issues with the air traffic control (ATC) system in the United States. H.R. 2997, introduced in 2017, aimed to transfer ATC functions out of the FAA to a separate nonprofit entity.[63] The United States is different in this way from many other countries. The FAA operates American ATC services, it is responsible for the governance of ATC, and it regulates safety in the industry. By comparison, other countries delegate these functions to separate agencies. Even the U.S. methodology of funding the flight process is different from the rest of the world. Whereas many major countries charge user fees, the United States relies on a series of taxes and general funding to support ATC operations.[64] These seemingly minor differences affect the ATC system's flexibility and capability to adjust with the times. For example, countries that have privatized or created alternative structures for ATC can raise capital through borrowing—unlike the FAA, which cannot.[65]

Centralized management of ATC in the United States has left U.S. ATC systems technologically behind their counterparts. The FAA still substantially manages flights by relying on paper slips in ATC towers, a system that is only now being phased out. Some U.S. airports will not go to electronic flight slips until 2025.[66]

When it comes to managing the low-altitude airspace where drones will ultimately operate, regulators need to view them as an opportunity to try something new and different in air traffic control. Attempting to maintain the status quo will take an industry bursting at the seams and weigh it down with a system known to be lethargic and disadvantaged compared to the rest of the world. Considering that the market for drones will be extremely valuable, the United States would be well advised to be a leader rather than a follower in this space.

POLICY SOLUTIONS

The United States prides itself on being a hub for innovation and views itself as a leader, pushing the envelope of what is possible. However, the country is not living up to this promise in its approach to drone technology. A combination of social pressures and antiquated regulatory hurdles has left the industry unable to operate at its full potential. The question is how to facilitate the necessary innovation while reasonably protecting consumers from unacceptable harm. If Congress and the FAA can strike the right balance, the drone industry will likely unleash a wave of innovation with the potential to profoundly impact the world. There are multiple policy options that the United States can and should pursue to live up to its promise of being the hub for innovation globally.

PERMISSIONLESS INNOVATION

In the realm of U.S. UAV policy, many regulations are probably unnecessary. Regulations on the federal and state levels have stunted growth in a sector that could significantly benefit consumers and producers. The safe delivery of medical supplies, protective equipment, and food could have been accomplished at a rapid pace if allowed to do so. Consider that federal, state, and local governments waived hundreds of regulations in the name of fighting COVID-19. Months have passed since governors initially waived some restrictions, and life has seemingly gone on without incident. The same could be true for UAV policy. The reality is that some of the rules on the books are not needed or could be modified.

A potential solution to some of these issues would be to embrace "permissionless innovation" principles and use a market empowerment framework to reimagine UAV policy. Rather than following the precautionary principle, preempting innovation in the sector in the process, policymakers and regulators should shape UAV rules and regulations to reflect the risks associated with the products in question.

Permissionless innovation is a theory of regulation that embraces the process of tinkering, experimenting, and exploring new ideas and ways of doing business. Permissionless innovation, as Adam Thierer of the Mercatus Center writes, is about freedom.[67] The United States has benefited from empowering this kind of freedom, leading to all kinds of advancements in technology and numerous industries. And Americans have gained from these advancements, benefiting from exposure to new products, cheaper services, more options, and increased welfare.

One way to think about permissionless innovation is the Market Empowerment Framework, a tool developed by the Libertas Institute.[68] The tool can help legislators, regulators, and individuals utilize the least restrictive means approach to identify the appropriate level of regulation a particular good or service may need. That, in turn, can help shape the country as a leader in regulatory reform.

THE MARKET EMPOWERMENT FRAMEWORK

Let's review the framework, the different components that make up each section, and the results of the different scores. Six core questions help shape how a legislator or regulator should approach a particular industry:

1. What is the good or service we want to analyze?
2. Is the use of the good or service commercial or public (government) purposes?
3. Can the good or service cause harm to consumers? If so, what kind (mental, financial, physical, etc.)?
4. What is the severity of the harm done to consumers (none, low, medium, high)?
5. What is the probability that the good or service will harm the consumer (none, low, medium, high)?
6. What is the permanence of the potential damages done to consumers (none, low, medium, high)?

In reviewing questions 4–6, the policymaker should select the option that best describes his or her belief. Each answer has a point value assigned to it:

1. None: 0
2. Low: 1
3. Medium: 2
4. High: 3

When all the points are tallied, the total score dictates the level of regulation appropriate for the good or service in question.

EXPLAINING THE QUESTIONS

Identifying the good or service for analysis

It might seem self-evident, but it bears stating that identifying the good or service is an essential step in the process. The good or service has to be identified to establish the accompanying mental framework and use the scorecard properly. Skipping this step can be extraordinarily problematic.

Use of the good or service

Think of this question as refining the lens the individual applies to the identified good or service. For example, let's say the idea is to study facial recognition technology, and in particular, commercial uses of the technology. That refinement results in an answer radically different from consideration of government uses of the same technology. The intended uses differ fundamentally depending on the case. The use case question serves as a central reminder in framing the necessary regulations or guardrails.

Identifying the harm

It is imperative to consider the harm a good or service can cause to a consumer. To begin, note that the term "harm" is too general. A more refined definition accounts for the types of harm, such as mental, physical, or financial harm, and potentially ranks them. For example, if a good or service potentially puts the consumer at risk of death, that is presumably more serious than other risks. Identifying harm allows a legislator or regulator to understand the good or service's most critical issues and tailor the focus accordingly.

The severity of the harm

Identifying the harm is important, but as noted earlier, understanding the severity of the harm can be equally important. If historical data suggest that a good or service's harm to consumers is modest, that information shifts the regulatory framework toward milder measures. When the data suggest the harm is significant, the appropriate level of regulation shifts according to the risk profile of the good or service. Understanding the severity of the harm can empower the legislator or regulator to embark on a more tailored path that appropriately accounts for risk instead of needlessly overstating risks, which regulators may otherwise be prone to do.

The probability of harm occurring

After identifying the type of harm and its severity, the legislator or regulator should consider the probability that the good or service will inflict that harm. For example, drones could potentially cause physical harm to a consumer. Maybe during a drone flight to deliver much-needed toilet paper to a consumer, a bird collides with the drone, causing it to crash and hit a bystander. Should that event unfold, it would be most unfortunate. But how likely is a drone to run into a bird? How likely is the drone to fall and cause damages? How likely is someone to die if hit by a falling drone? The available data will influence the score. The likelihood of harm occurring shapes potential policy prescriptions as a result.

The permanence of the harm

Some things are easier to fix than others. The difficulty or ease of making a harmed consumer whole is a critical factor. Solving a dispute involving a tree on the border of two properties can be easy. However, if a fully autonomous car gets into an accident, the issues that arise can be significantly more complicated. In thinking through the permanence of the harm, the focus should be on the difficulty of making a person whole after being harmed.

THE SCORING PROCESS

After answering all six questions and scoring questions 4–6, the legislator or regulator can tally up the total points earned by the good or service being reviewed. Whatever the total amount is, it equates to the least restrictive means of implementing rules and regulations on that good or service. The policy prescriptions, depending on the score, are as follows.

0: Permissionless innovation

Permissionless innovation is the principle that innovators do not require any regulatory framework to begin the process of tinkering, experimenting, and exploring new ideas and ways of doing business. It is the least restrictive approach to regulating companies.

1: Social norms and pressures

Under this model, a company can adapt its business practices as consumer behavior and demands change over time. This practice is prevalent in many leading businesses in the United States today. For example, in 2017, American fast-food chain McDonald's decided to switch from plastic straws to paper straws in its stores in the United Kingdom. It made this decision because it had a sizable number of customers who wanted to see that change occur.[69] The company could have waited for regulations or legislation to be enforced, but it voluntarily went down that route because it felt the company would ultimately benefit from the change because it listened to its customers. Businesses almost always want to use this strategy because it benefits their company in the long run.

2: Learning/coping/experience/experiments

Sometimes a business may not know what is right for its market. It may choose to bring out a new product or service to try new things and then adjust accordingly. Because this score indicates the product or service presents low risk, allowing businesses to experiment within a regulatory environment is sufficient. These kinds of companies might be ideally suited to participate in regulatory sandbox programs, in which regulators supervise business innovation while temporarily freezing regulations and penalties. (See a more thorough discussion in the section, "Expanding Participation in an Industry-Specific Regulatory Sandbox for Drones.")

3: Self-regulation

Under this model, companies can establish their own rules and best practices guided by consumer behavior and market pressure. Companies are aware of the potential risks and harms associated with their product. They lead the way in formulating best practices and self-regulatory systems based on their intimate knowledge of their product. Companies can employ various tools, such as creating a code of conduct and expectations or establishing standard operating procedures and protocols. This method is relatively malleable, as the company is responsive to a changing environment, new social concerns, emerging technologies, and other factors that may arise, and can quickly adapt as needed.

4: Education/media literacy

This is the preferred method of regulation for companies with more complicated products than a standard consumer good (e.g., food, drinks, banks). This model gives businesses the opportunity to educate the public about their products. For example, financial technology companies such as Robinhood—which encourages investment in the market with any type of currency, with the added benefit of a zero-commission-based structure for transactions—have had to promote media literacy so people could better understand the product and how it worked. Robinhood's goal was to make investing as easy as humanly possible for everyday consumers; that meant not only creating the product itself, but also spending significant time and energy creating the educational materials to make consumers comfortable using the product to make trades.

5: Labeling/transparency

Goods and services in this category might cause some degree of harm, so consumers should be particularly informed about that good or service. Direct information via labeling or other transparent communications can suffice. For example, labeling can help consumers understand the quality and cut of certain meats and provide contextual information they may otherwise not have. This process is likewise helpful for pharmaceutical drugs so consumers can make informed choices. It provides a way to adequately inform consumers and mitigate potential risks associated with certain low-to-medium risk goods and services.

6: Industry guidance

As a particular industry grows, there may be a desire to standardize the best business practices and industry methods. Under this style of regulation, the businesses within an industry agree to a set of rules that everyone will follow. For example, manufacturers of a specific good might want to establish guidelines on everything from production processes to the work environment. In standardizing, the various manufacturers self-police and adhere to a particular set of guiding principles.

7: Licensing and permits

Licensing and permits are a top-down style of regulation. Typically, regulators favor licensing and permits for goods and services that require extensive training to understand and master. Restricting the ability to use or operate a particular product may be necessary when consumers may not know how to use the product. For example, someone responsible for designing a building may want to hire a licensed architect because the building will ultimately have numerous people inside. If poorly designed, all those people would potentially be needlessly at risk.

8: Administrative mandate

Like licensing and permits, administrative mandates are top-down regulation styles in which the state essentially sets the rules by which a business can operate within the state. From enforcing environmental guidelines to requiring specific certifications or special insurance, the government will typically not allow a company to operate without satisfying all its concerns.

9: Precautionary principle

The precautionary principle is a regulatory lens that places an extreme emphasis on protecting the consumer. While the precautionary principle has many variations, Cass Sunstein says they all share a common feature: promoting the idea that "regulators should take steps to protect against potential harms, even if causal chains are unclear and even if we do not know if those harms will come to fruition."[70] This type of regulation is by far the most top-down restrictive style. In these cases, the government is extraordinarily protective of consumers and, as a result, is very heavy-handed in regulating how specific companies participate in their markets, if it allows them at all.

THE FRAMEWORK IN SUM

As the score increases from 0 to 9, the policy prescription is increasingly precautionary. The prescriptions from 0 to 2 may be described as emphasizing *adaptability*. The focus is on the firm being able to adapt to changing environments and address concerns flexibly and quickly.

From 3 to 5, the policy prescription turns to *resiliency*. It relies on slightly more targeted solutions for the particular good or service to address the potential harms it may cause.

Scores in the range of 6 to 8 indicate the need for *anticipatory regulation*. The policy prescriptions in this range restrict the ability to utilize a particular good or service in one way or another.

Anything that receives a score of 9 falls into the category of *prohibition*. Typically for these kinds of goods or services, not enough information is available to make an informed decision about the necessary level of regulation. Regulators often apply precautionary principles to products in this category, subjecting them to numerous stringent measures, from outright product bans to censorship, or suppressing information on them from the general public. Further research and development are needed to answer some of the questions posed in the future.

This framework and scorecard empower legislators, regulators, and others to better understand how to answer questions about UAVs.

Focusing on the least restrictive means approach to regulation in the UAV industry enables the sector to be nimbler and respond more efficiently to consumers' needs and concerns. In so doing, consumers receive the numerous benefits that the UAV industry can offer.

ELIMINATING LINE OF SIGHT REQUIREMENTS FOR DRONES

The FAA defines the line of sight as "vision that is unaided by any device other than corrective lenses" with which "the remote pilot in command, the visual observer (if one is used), and the person manipulating the flight control of the small unmanned aircraft system must be able to see the unmanned aircraft throughout the entire flight."[71] The stated purpose of this regulation is to ensure that operators

1. know the location of the unmanned aircraft;
2. determine the unmanned aircraft's attitude, altitude, and direction of flight;
3. observe the airspace for other air traffic or hazards; and
4. determine that the unmanned aircraft does not threaten the life or property of another.[72]

Current technologies can address many of the issues the FAA wants to solve without retaining the mandate that an unmanned aircraft remain within the visual line of sight. The agency is overestimating the risk associated with operating commercial unmanned aircraft given new technologies that make such operation significantly safer.

For example, DJI, one of the leading drone manufacturers, recently released an application for its drone software that would inform users of the altitude, speed, and direction of flight. According to the company, "Using a simple app, anyone within radio range of the drone can receive that signal and learn the location, altitude, speed and direction of the drone, as well as an identification number for the drone and the location of the pilot."[73]

DJI's drones also address the third concern of the visual line of sight requirement—observing airspace for other traffic and hazards—by equipping each drone with dozens of sensors. Thus numerous redundancies are in place to assist in the event something fails.[74] The drone's algorithms identify objects in its flight path and adjust accordingly. Implementing this technology within the drone itself provides a pathway to establishing skies that UAVs can navigate safely.[75]

A June 2018 report from the National Academies of Sciences highlighted this issue. The committee that wrote the report expressed the belief that the FAA was unevenly applying the types of risk associated with airplanes to unmanned aircraft. Gregory Baecher, one of the committee members, stated that the FAA with its "near-zero tolerance of risk is stifling innovation and putting an unnecessary burden on the [unmanned aircraft system (UAS)] industry. There needs to be a shift towards a more holistic decision-making process that weighs risks against benefits."[76]

The committee also recommended that the FAA change its processes and requested that the agency commit to reviewing its risk assessment standards within six months of the report's publication.[77] Congress seemingly agreed. The FAA Authorization Act of 2018 directed the agency to address its risk assessment methodology and adopt a risk-based approach when considering drone policy.[78]

With the elimination of line of sight requirements, drones will finally have an opportunity to test their limits. The use of drones across various industries provides numerous benefits *and* is safer than putting humans into potentially risky situations.

ENCOURAGING BOTTOM-UP DEVELOPMENT OF AIR HIGHWAYS

The FAA must come to understand that it cannot control every inch of airspace. While it is certainly expected that the agency be responsible for interstate air activity with manned aircraft, extending that control down to heights or locations where planes do not reasonably operate is simply not practical. The FAA should allow states to develop and manage aerial corridors for low-altitude flights. That way, states can create pathways for drone operations that do not interfere with FAA operations in a given area.

An aerial corridor is a designated portion of airspace where an aircraft (in this case, UAV) must stay during its transit in a given area. One of the main points in designing aerial corridors is the need for clear demarcation of different types of aircraft. This step is essential to ensure that commercial drones are not merely roaming free in the sky. It is therefore critical for policymakers and state regulators to design aerial corridors in a way that avoids airports, schools, homes, or other sensitive areas.[79]

As long as drones accurately follow their routes within an aerial corridor, the issues will be minimal. Suppose a drone deviates from its designated aerial corridor for some reason other than to deliver a product.

In that case, the company responsible for the drone can be held accountable for violating the rules through various enforcement mechanisms. Deviations are already easily verifiable through GPS technology, which is increasingly incorporated in drones.

States need to be responsible for creating and managing air highways in low-altitude space within their state. States possess the local knowledge necessary to build an intrastate aerial corridor network that best suits their needs and addresses their concerns. No doubt, states would welcome the FAA's guidance and expertise on how corridors should look—offered in an advisory capacity. But the FAA should not be the primary driver of efforts to build out aerial corridors within states. Allowing states to run the drone highways means they can set rules and limitations on time, manner, place, and noise and address privacy concerns in a more tailored manner.[80]

As an added benefit, states that manage their air highways could generate passive income for the state by either leasing or auctioning access to the public right-of-way between 50 and 200 feet above ground level.[81] With the number of industries that have an explicit interest in the technology, that could be an extremely lucrative endeavor. Another benefit of states creating their air highways is that they could run the aerial corridor over publicly owned property. Doing so would avoid legal issues that might arise if corridors extended over private property. Ideally, air highways would run above state and local roads, to open up as much of the state as humanly possible to reap the benefits that commercial drones offer.[82] It is worth noting that in states creating and managing air highways, nothing would (or should) prevent public entities or agencies from taking advantage of these air highways for public use within reason.[83]

The central question in the creation and management of air highways pertains to state preemption, a legal doctrine that allows state law to prevail when at odds with local law. Will states use preemption to prevent local authorities from taking action against commercial drones? State preemption would be ideal; if localities were to set individual rules and regulations on the use of commercial drones, the wide array of possible policy outcomes could make compliance hard to achieve. Commercial drone companies would likely face hundreds, if not thousands, of unique regulations within a state, to say nothing of restrictions in the other 49—or worldwide.

For all of these reasons, the FAA should establish a process for states to develop and manage air highways. Doing so would allow the agency to shift its attention to other policy areas. The FAA can and possibly should work with stakeholders to establish specific standards at the federal level related to interstate usage of commercial drones. But the agency should take a hands-off approach and simply offer its expertise in helping states get their air highway projects up and running.

ESTABLISHING AIR RIGHTS AS A PATH TO LEGITIMACY

States also need to establish air rights to manage their drone highway programs effectively. "Air right" is defined as "a right of way in the air space above a property owner's land and the immovable property on it, subject to the public right of air navigation above the property at a legally prescribed altitude."[84] One of the benefits of establishing air rights is that they explicitly define property rights in the state.[85] Another benefit is that they essentially grant legitimacy to the air highway programs established by the state.

While air rights have historically been associated with real estate development within cities, they are increasingly relevant for drone routes.[86] Thus, in establishing these rights, everyone wins; the consumer, the homeowner, and the commercial entities using the technology all benefit. A rights framework creates the rules and expectations that everyone abides by, while simultaneously protecting companies from potentially frivolous lawsuits resulting from a lack of clarity about property rights above a person's home.[87]

AIR NAVIGATION EASEMENTS

Air navigation easements—sometimes called avigation easements—grant permission to fly over one's land. They convey property rights in specific instances but do not give ownership.[88] Airports already use them to ensure airplanes can take off and land safely.

While air property rights are useful in establishing certain fundamental rules, the existing rules about trespassing remain in force when a drone descends to make a delivery. As drones' commercial and recreational use continues to rise, so will the inherent conflict between landowners and drone users. Avigation easements can offer solutions that ultimately benefit all parties in the long haul. The commercial drone industry will need some form of avigation easement because operators are likely violating trespassing laws when delivering packages. With an easement, so long as the drones are high enough not to bother residents, drones could operate with relative ease.[89] The state would play a role in granting avigation easements by establishing a flexible aerial corridor. If states take advantage of the opportunity to demarcate existing airways and public roads, a drone would be able to seamlessly transition from an air highway to an avigation easement as it makes its delivery and returns to its point of origin. Companies would not have to worry about not-in-my-backyard (NIMBY) individuals who might want to bring a lawsuit, as the easement would allow corporations to perform their duty without issue. Or, put another way, the avigation easement would deter NIMBY homeowners from pursuing frivolous lawsuits against companies using commercial drones.

An avigation easement would not render a drone operator immune from civil liability. If a consumer faced harm resulting from an incident with a drone, the operator or owner could undoubtedly face civil liability lawsuits for issues stemming from damages the drone may have inflicted.

SEPARATION OF DRONE AND STATE

When discussing UAV policy, policymakers must consider who is trying to leverage the technology and what their intentions are. Regulations need to be tailored because the intended use varies from one domain to another.

When drones are being used in the private sector, regulators should presume, in the absence of compelling evidence to the contrary, that it's not for nefarious purposes. When a company wants to use a UAV, it is almost certainly more efficient than a conventional vehicle, which is a huge benefit in itself. This technology should be cheered and encouraged, as it represents great ingenuity and creativity.

Entrepreneurs need to be encouraged and rewarded for taking risks that can potentially improve countless people's lives. And even if their idea does not work out, they will have contributed to the development of the technology. People can learn what does and does not work and modify accordingly based on their own failure or that of others.

Private companies are highly responsive to social pressure. As a result, they are more inclined to take precautionary measures to avert potential issues with consumers. The legal system can sort through those complaints. And bad actors won't last long, as consumers will stop using goods and services from companies that violate their trust.

The situation is different for government use. When the government wants to use the same technology, it does not face all of the same constraints, and government's greater ability to harm citizens means that policymakers need to press the pause button. Under the Market Empowerment Framework, it might be desirable to intentionally place more stringent requirements on the government's ability to use such technologies out of concerns for abuse and misuse of the technology. Among the primary ways governments use drone technology are surveillance and reconnaissance. The military relies on drones both for their surveillance capacity and as a weapon of war, and significant reform in that arena will be hard to achieve, barring strong congressional or presidential action. Perhaps a more viable option is to promote transparency, encouraging the government to inform the public about how the military uses drones abroad as well as the monetary and nonmonetary costs associated with their use. Deploying drones for strikes and surveillance should not become normal behavior. Such action should bother people to their core because lives hang in the balance of a limited number of people's decisions. Striving for transparency can go a long way in establishing norms and holding accountable those who abuse the technology.

Law enforcement agencies' use of drones inside the United States should raise red flags. The government's use of drones to monitor protests during the coronavirus lockdowns is clearly a threat to people's constitutionally protected right to assemble peacefully.

In another setting, it was discovered that a surveillance plane had flown over the city of Baltimore as part of a pilot program. A city employee, speaking for himself, responded, "Having a tool that the police can use that doesn't really add any additional surveillance, it just gives them the option to do it, isn't really a horrible infringement on our rights because nobody expects privacy outside anymore."[90] But wanting to solve crimes—by no means a bad thing—cannot come at the expense of, and right to, privacy.

The end-use is also worth noting. When one person films another without consent and uploads the video to a website, there are mechanisms in place to take it down. The harm is drastically different when law enforcement uses the same technology. Time and time again, law enforcement has utilized various forms of technology then ultimately been found to be arresting innocent people. The reasonable expectation of the right to privacy means that Americans should not be subjected to an Orwellian surveillance state.[91]

When a government entity wants to leverage technologies like UAVs, guardrails must be in place to ensure that it is not actively violating individuals' civil liberties. An audit process is one method of holding law enforcement agents accountable when they use technology to solve crimes.

Drones, like most technologies, have uses that can be both positive and negative. Technology can serve as a powerful tool in the kit for law enforcement to help solve crimes—which is excellent. However, this use of technology cannot come at the expense of individuals' fundamentally guaranteed rights.

In Utah during the 2021 general session, the state legislature passed House Bill 243, Privacy Protection Amendments, in an effort to tackle some of these critical issues.[92] The bill creates two "privacy officer" positions, one housed within the governor's office to deal with executive agencies, and the second within the state auditor's office to address non-executive government entities. They will be responsible for handling the civil liberty and privacy questions related to certain government technology uses. The officers, working with an advisory committee composed of numerous experts in privacy, technology, and law enforcement, will evaluate technologies and inform the public about the government uses and existing privacy concerns.

The legislation is a first-in-the-nation endeavor that closely followed the Privacy Protection Act proposal, which sought to strike a better balance between public safety considerations and an individual's reasonable expectation of and right to privacy.[93] It creates a system that establishes public buy-in, holds government agents accountable when they inappropriately leverage technology, and proactively establishes guardrails to protect civil liberties, rather than waiting years for courts to resolve any outstanding issues. Supreme Court Justice Samuel Alito recognized the problem of relying on the court to solve these issues, stating, "[Courts] are very ill-positioned to make these determinations. . . . We are not up on all the latest technology. If privacy is to be protected in the future . . . state legislatures should take the lead."[94]

State legislatures can take the lead when looking to establish UAV rules by addressing commercial uses separately from public uses by government entities. Doing so reinforces how different uses of UAVs can have drastically different positive and negative impacts on American lives.

EXPANDING PARTICIPATION IN AN INDUSTRY-SPECIFIC REGULATORY SANDBOX FOR DRONES

States and the federal government should begin by acknowledging that they likely do not have the right set of regulations for UAVs. States should look to place the industry in a regulatory sandbox for a limited period, and the FAA should expand its existing drone sandbox program to include more states.

In October 2017, a presidential memorandum established the Unmanned Aircraft Systems Integration Pilot Program (IPP). Run through the Department of Transportation (DOT), the program enabled state, local, and tribal governments to work with private industry to advance drone operations in the National Airspace System (NAS). The program facilitated a productive conversation, attempting to strike a balance between state and national interests on drone integration, and provided DOT some actionable information for how to universalize and expand drone integration into the NAS. The program officially came to a close in October 2020 by an administrative mandate.[95]

The FAA found the program useful and immediately opened a new program in October 2020 called BEYOND. Its purpose is to work with eight of the original nine participants from the IPP to tackle issues highlighted by the agency in the integration of drones:

- Identify viable "beyond the visual line of sight" (BVLOS) operations that are repeatable and scalable.
- Use data gathered by participants to better understand the benefits drones can provide.
- Work with communities to identify their primary concerns and potential solutions.[96]

These FAA programs can be a great resource for states pursuing this process. States can see how the technology can work for them without getting into turf wars with the federal government. These regulatory sandbox-style programs are a great way of facilitating the necessary interaction and communication between all parties while simultaneously advancing a promising technology.

State regulators, who possess more local knowledge of potential issues in their jurisdiction than their federal counterparts, can achieve a more dynamic and flexible regulatory environment by working hand in hand with industry stakeholders. Doing so allows for a more bottom-up approach to dealing with issues associated with UAVs pragmatically, rather than forcing a square peg into a round hole.

A regulatory sandbox is a unique legal classification that creates space for regulators to freeze regulations and penalties temporarily. The process allows private companies to develop or introduce an innovative product or service into a market space where current industry standards do not exist yet or apply to their product.

A regulatory sandbox can be molded in two distinct ways, industry specific or industry agnostic, depending on how legislators and regulators want to pursue innovation. A significant common feature is that a regulatory sandbox facilitates the necessary dialogue between market participants and regulators, which in turn informs regulatory actions so that they strike the right balance between promoting innovation and mitigating potential risks.

Regulatory sandboxes originated in 2014 in the United Kingdom at the Financial Conduct Authority as part of a larger initiative called Project Innovate. Project Innovate aimed to create competition for the betterment of the consumer. Those involved wanted to fix concrete problems facing consumers in finance via financial inclusion and flexible pilot testing programs. As a result, the first regulatory sandbox was specifically geared toward financial technology (fintech) companies.[97] Experiencing tremendous success with its early cohorts in 2018, the Financial Conduct Authority announced that it was expanding its sandbox to an international scale, allowing companies from all over the world to apply.[98]

Regulatory sandboxes have since been launched around the world in places such as Abu Dhabi, Canada, Denmark, and Hong Kong. Singapore, the second country to formally launch a fintech sandbox, took a formal "never say no approach," encouraging its regulatory agencies to allow piloting to happen, even in industries that are typically subject to tight regulation.

South Korea went a step further by creating a general sandbox. The goal was to provide local companies with more freedom rather than preemptive regulation. The government set up the broad sandbox to encourage new startups and foster new economic growth.[99] The South Korean regulatory sandbox aimed to pave the way for success by having companies launch new goods and services first and retroactively apply reasonable regulations later.

In 2018, Arizona became the first state to launch a fintech sandbox, allowing companies to remain in the sandbox for up to two years and service up to 10,000 clients before applying for a formal license.[100]

In 2019, Utah became the third state to launch a fintech sandbox, allowing companies to test new and innovative ideas in financial products and services without being licensed. Additionally, in August 2019, Utah's Supreme Court approved a pilot program allowing for nontraditional legal services. The court supported this measure to "profoundly reimagin[e] the way legal services are regulated in order to harness the power of entrepreneurship, capital, and machine learning in the legal arena."[101] One of the potential reforms for this "sandbox" is enabling non-practicing lawyers expanded scope of practice and even ownership in law firms.[102] On September 8, 2020, the Utah Supreme Court announced it accepted five participants into the legal services sandbox, ranging from large firms such as Rocket Lawyer to smaller companies such as 1Law.[103]

The U.S. government has also experimented with sandboxes. In 2017, the FAA set up a drone sandbox to pair state, local, and federal regulators with private actors to work on drone integration.[104] In the FAA Modernization and Reform Act of 2012, Congress instructed the FAA to create six test sites for UAS technology and incorporate it into the NAS over five years.[105] The program became operational in 2014, with test sites in Alaska, Arkansas, Nevada, New Mexico, New York, Texas, and Virginia, when test flights commenced.[106] While it was positive to see the FAA admit more states than Congress requested, the number of test sites is still underwhelming. The sandbox program needs to expand significantly and allow all states to participate. More testing and experimentation can lead to further discovery and development of the technology, more innovations, and improvements in the UAS industry.

Flexibility and adaptability are crucial in facilitating effective regulatory regimes. As Rob Morgan of the American Bankers Association said, "You're only as innovative as your least innovative regulator."[107] A significant benefit of being inside a sandbox is the exemption from enforcement if a company violates regulations that would typically require the business to cease its legally complex operations. However, regulators need to be able to address the valid concerns raised about sandboxes. Most notably, they need to balance promoting innovation and protecting consumers.

The primary concern, of course, is consumer protection. How can a sandbox program be structured to ensure a company does not victimize the consumer through its participation in the sandbox? Another concern is the process of selecting an individual to lead the program, and how that person or the program will hold participating companies accountable. Another problem lies in the stability of the program itself. What steps should policymakers take to ensure the long-term success of the sandbox program? Furthermore, concerns have been raised about the number of consumers exposed to fringe and frontier technologies; how can the risk be minimized in these circumstances?

As with any government program, one challenge regulatory sandboxes face is how to deal with questions surrounding regulatory capture and crony capitalism that could develop within the program. Crony capitalism is "an economic system where businesses and individuals with political connections and influence are favored in ways that are seen as suppressing open competition in a free market."[108] Theoretically, larger firms with such connections could abuse these programs by designing regulations that could shut out potential competition.

Regulatory sandbox programs, like most public institutions, rely on establishing buy-in and trust from the public. Should a case of cronyism arise, it would undermine the program's credibility and make it increasingly challenging to achieve the stated goal of promoting innovation.

Notably, to this point, states that have implemented regulatory sandboxes have attracted a wide array of firm sizes. For example, in Arizona's fintech sandbox, most of the companies employ fewer than 10 people. Additionally, because they are required to demonstrate that the public benefits from the companies' products, many of the participants offer products targeted at helping the state's residents. From income-sharing agreements to car loan refinancing to other innovative business solutions, sandbox participants play a role in solving problems faced by Arizona residents.[109] The state's fintech sandbox was started in 2018, and there have been no reports of cronyism or corruption to date.

Looking at Utah's legal services sandbox, the Office of Legal Services Innovation released a report detailing the participants in its program, including information on the number of lawyers working in different fields and the different governance structures employed by participants. As of November 2020, the office had accepted 12 applicants into the sandbox. The office received over 33 applications at a rate of nearly two a week. There is a diversity in firm size, the type of law practiced, and the category of risk associated with each applicant. This particular sandbox program highlights the office's restraint (not just letting anyone in) and the diverse background of participants, which demonstrate that firms of all sizes are taking part in the attempt to innovate an otherwise lethargic industry.[110]

While there is no silver bullet that can truly solve the problems surrounding regulatory capture and crony capitalism, institutional safeguards can mitigate the propensity for such incidents to occur. For example, requiring that the program director be appointed by the governor and confirmed by the Senate could ensure that a candidate for the office is more likely to carry out its mission than a purely political appointee might be. Another option would be to require that notification of an applicant's acceptance into the sandbox be published within the pertinent industry, to allow competitors to apply for the same regulatory relief if they so choose.

Regulatory sandboxes as tools for dynamic regulatory reform and promoting innovation are still relatively new institutions that numerous states are considering using. As they gain popularity across the country, it will be important to monitor their progress and make necessary changes to curb potential inroads for cronyism.

Finally, there are concerns about federal preemption. How can a state position itself to be successful, facilitate innovation, and protect consumers while federal programs may prevent them from doing so? Legislators can help regulators strike this balance by focusing on five core areas in policy:

Transparency and the buyer beware doctrine. Companies that participate in the sandbox need to be clear about the product they are offering consumers. If a company is disingenuous about what it is trying to do, that will dissolve trust, undermining the program.

Accountability. When a firm has harmed consumers, it is crucial to take steps to make consumers whole. If there is no accountability, then the institution will be weakened.

Institutional soundness. A challenge facing sandboxes is that the regulatory institutions managing the sandboxes have to be supported. They need to be well funded, and they need to be able to optimize the lessons learned from previous cohorts and adapt their practices for future cohorts.[111]

Exposure control. It is essential to mitigate the risk of exposure to consumers from goods and services that sandbox participants offer. If an incident occurs, it can be more easily resolved in a smaller, controlled population.

Federal interaction. Many companies in a state-level sandbox may also want to be involved with similar programs offered on the federal level. Legislators worry about preemption potentially interfering with their responsibility to protect their citizens, so working with the respective federal agencies is imperative to find common ground.

An additional method of protecting the consumer is to require the businesses in the sandbox to take out some extra liability protection (within reason). Doing so mitigates the risk of consumers, if harmed, finding themselves unable to get restitution for the harm done to them by a participating company. If a company causes significant harm to consumers, the regulator can remove the company responsible from the sandbox altogether. Accountability measures increase credibility in the system and are also a potential barrier against bad actors who might wish to exploit the sandbox.

Regulation and fear of the unknown should not impair innovation. The United States has traditionally been viewed as a haven for innovation and progress, paving the way with the Industrial Revolution and transitioning to the age of technology. Giving in to technophobic tendencies that stifle and restrain innovation will have negative impacts that could last for decades. Continuing to pursue prescriptive policies in an increasingly dynamic and competitive global market will indeed chase companies out of the country to more friendly environments.

CONCLUSION

The future of drones looks extremely promising. The technology can offer significant benefits to consumers, producers, and governments alike. Yet policymakers have various questions to consider.

By following the principles of permissionless innovation and a least-restrictive-means approach to regulation, the drone industry can reach its fullest potential. Drones are not airplanes, and the FAA's treatment thus far has been detrimental to the industry's ability to innovate and develop. Reforming the agency's risk assessment is a critical step in fundamentally changing the way drones are perceived and regulated going forward.

Delegating the creation and management of air highways to the states offers the FAA much-needed relief to focus on the big-picture issues while simultaneously offering local agencies a unique and newfound way to experiment and earn passive income.

By establishing air rights and pushing for avigation easements, firms can safely operate without fearing potential litigation. The drone sandbox program in the United States should be more robust than it currently is, which can be achieved by expanding access to more states. With greater participation in the program, companies can more rapidly develop the technology and adjust as needed—as opposed to the piecemeal approach of slowly letting companies and states experiment with the technology. There is no justification for making Amazon's Prime Air service wait seven years for permission to use the same technology the company had been using successfully overseas.[112]

It is also crucial to have a conversation, at both the state and federal level, about steps policymakers can take to mitigate Americans' concerns about the expectation and right to privacy.

In implementing most, or even some, of these critical policy changes, America stands to make good on its promise of being a hub of innovation. Then the country can genuinely say it's a place where entrepreneurs can experiment and the citizens benefit from unlocking human ingenuity. Technological anxieties will persist among many Americans, but that should not deter the country as it looks forward. The United States has traditionally striven not to shy away out of fear but rather to embrace new technologies, take on whatever challenges may arise, and rise above them. Embracing the numerous benefits UAVs have to offer will play a key role in allowing the country to reach new heights.

CHAPTER FOUR

WHO SHOULD GOVERN THE SKIES?

BRENT SKORUP

Commercial drone technology has improved rapidly in the past 10 years. The drone market circa 2010 was primarily expensive weapons of war, but today manufacturers are making drones for hobbyists and for major industries, including medicine, logistics, telecommunications, and public safety. The primary challenges to mass-market drone services in the United States are regulatory rather than technological.[1]

For instance, in a 2019 interview, economist Tyler Cowen put his finger on the difficult property rights and jurisdiction issues raised by drone flights: "How are we going to have the easements for the air, where do the property rights really lie? . . . It will take quite a while to untangle that mess."[2] A reporter asked the CEO of Zipline, a U.S.-based drone services company that has completed tens of thousands of medical deliveries in Ghana and Rwanda, why Zipline wasn't delivering in the United States. CEO Keller Rinaudo pointed to regulatory obstacles. "In the U.S. there's this sense that this technology is impossible, whereas it's already operating at multi-national scale, serving thousands of hospitals and health facilities, and it's completely boring to the people who are benefiting from it."[3]

Advanced drone technology is at the stage that operators can deploy long-distance (that is, beyond the visual line of sight) services quickly—as soon as the government grants them access to airspace. In April 2020, the United Parcel Service (UPS) and drone operator DroneUp revealed the speed at which drone delivery can commence once they have regulatory approval to use low-altitude airspace. The companies had a unique opportunity to test services on an empty college campus in Virginia:

> "DroneUp and UPS did the most extensive delivery of packages that has ever been done [by drones]," says Tom Walker, DroneUp CEO. "Hundreds, if not thousands of flights [completed]—it was an exhaustive exercise. We took a [vacant] 55-acre college campus, we made it a town, and by the end of day two we were doing deliveries every 3 minutes."[4]

Despite the cautious approach from regulators, drone technology advancement and demand for drone services exceed even expert predictions. U.S. regulators have consistently underestimated the growth of the commercial drone market. In 2014, the Federal Aviation Administration (FAA) estimated there would be 7,500 commercial drones in 2018.[5] The actual number in 2018 was more than 110,000 commercial drones.[6] In 2018, the FAA made a new projection: in 2022 commercial drones would number 450,000.[7] By 2020 that number had been exceeded—two years ahead of FAA projections.[8]

This "pacing problem" is common in technology fields.[9] Regulation, especially in established industries like aviation, cannot keep up when technology shocks and regulatory bottlenecks emerge. Airspace access is the primary bottleneck for small drone companies, and soon large passenger and freight drone companies will run into the same obstacle. Fortunately, U.S. aviation regulators (the FAA) and NASA are aware of the pacing problem for drones. As one NASA official conceded, "Current [aviation regulation] practices are too cumbersome to accommodate the 800,000-plus small drones that are expected by 2023. . . ."[10]

Congress instructed the FAA to integrate small drones into the National Airspace System a few years ago. To have an extensive commercial drone industry and passenger drone industry, there need to be drone highways—aerial corridors—crisscrossing towns, suburbs, and cities.

However, if the FAA were to extend these drone corridors unilaterally, it would face opposition not only from landowners but from state governments, which have a plausible claim of sovereignty and police powers over low-altitude airspace tied closely to the surface beneath it (hereafter, surface airspace). Many states expressly asserted sovereignty to surface airspace decades ago and are beginning to regulate that airspace for drones. Several states, for instance, have created no-fly zones for drones over sensitive areas, including critical infrastructure, schools, sports venues, and prisons.[11] Some cities likewise are prohibiting drone flights at low altitudes.[12]

National policy regarding how federal, state, and local governments will regulate drone airspace access is as confusing today as it was when the Congressional Research Service reported on the issue to Congress in 2013.[13] The FAA has accrued significant authority over aviation and airspace regulation since the 1950s, and many at the agency and throughout federal government resist the prospect of sharing authority with state governments. Major drone operators assert that only federal regulators have authority to define where drones operate and where they are prohibited.[14] In particular, they argue that drone no-fly zones "may only be established . . . by the federal government."[15] As explained later in "Airspace as Property and State Powers in Airspace Policy," this latter view is likely wrong, but the altitude at which state and property owners' powers are extinguished remains unclear. In the meantime, states, cities, and landowners are losing their patience as drones proliferate, and they are responding with laws, regulations, and litigation.

Quietly, coalitions are forming and preparing for legal battle. On the side of unitary federal authority are many people within the FAA and Congress and among the major drone operators and trade associations who fear a fragmented and chaotic regulatory environment if state airspace regulations proliferate. On the other side are state aviation officials, the real estate bar, grassroots groups, and small drone operators who want state and local authorities to exercise their powers and respond to local concerns that far-off federal regulators cannot accommodate.

Lawmakers must clarify two distinct questions. First, *who* will regulate drone airspace? In particular, will drone airspace jurisdiction be federal-centric like traditional aviation, or will jurisdiction be shared by federal, state, and local authorities under cooperative federalism, much like telecommunications and roadways? Second, *how* will airspace access be rationed? Will routes and terminals be administratively assigned like traditional aviation or allocated via market processes like radio spectrum and offshore oil sites?

This chapter argues that the answer to the first question, *who*, should be shared authority between federal and state transportation officials. Further, as for the second question, *how*, aviation regulators should create drone airspace markets. As explained later in "Airspace as Property and State Powers in Airspace Policy," absent a revolution in constitutional law, a cooperative federalism model is likely to emerge for drone airspace regulation. This conclusion is derived from the body of airspace law and earlier aviation cases, including the 1946 Supreme Court case *United States v. Causby,* dealing with invasions of homeowners' airspace. Further, the administrative rationing of airspace, routes, and slots in traditional aviation has hindered competition and innovation. To give new entrants opportunities in the drone delivery industry and to prevent damaging technology lock-in, regulators should consider demarcating and leasing aerial corridors to drone operators. The proposal to auction or lease airspace has received brief discussion by the FAA's Drone Advisory Committee,[16] Government Accountability Office (GAO) reports,[17] RAND,[18] and Airbus researchers,[19] but a fuller examination of the legal and practical realities for airspace leasing is included later.

FUTURE LONG-DISTANCE DRONE SERVICES

Most social value of drone operations is in long-distance and automated services. Making deliveries, shipping time-sensitive freight, or carrying passengers in the air makes entirely new markets, services, and industries possible. Automation of drone flights means huge productivity gains as a relatively small number of people can manage and operate extensive drone logistics. Automation also means that passenger drones one day will have no pilot cockpit, which improves air taxi economics and, one day, safety. These logistics and automation technologies are being tested around the world today and maturing rapidly because airspace—unlike roadways—is relatively empty, which makes automated technologies much simpler to test and achieve.

Drone delivery technology is maturing quickly. Companies such as CVS, Google, UPS, Walmart, and Zipline are creating or partnering with drone services companies to make residential and commercial deliveries of goods and supplies. In 2017, the White House and the FAA permitted about 10 drone programs throughout the United States, including some delivery services. Operators have been allowed higher-complexity operations over the years. One program in North Carolina overseen by the state's Department of Transportation completes medical deliveries, including the first long-distance drone delivery pilot program which began in the summer of 2020.

However, while pilot programs allow companies to test proof-of-concept, a sustainable business model cannot depend on regulatory waivers and temporary programs. Long-distance drone operations require airspace access and, in particular, widespread and liberal access to residences, commercial properties, and medical and corporate campuses. Small drones typically fly at less than 500 feet above ground level, so they pose a *de minimis* risk of collision with small, manned aircraft and helicopters. However, as explained later in "Airspace as Property and State Powers in Airspace Policy," flights at low altitudes raise trespass, nuisance, and takings issues from landowners, and regulators cannot simply authorize drone flights to any unoccupied airspace.

MODELS OF AIRSPACE REGULATION

The two major dimensions of public resources regulation to explore for low-altitude airspace regulation are *jurisdiction*—who regulates?—and *allocation*—how is the resource assigned among competing potential users? The primary alternatives of jurisdiction are federal authority and state authority. The primary alternatives of allocation are administrative allocation and market allocation. Natural resources and public resources tend to fit into one of these four categories. High-altitude airspace in traditional aviation is federal-administrative. Radio spectrum is federal-market. Appropriative water rights in the western United States are state-administrative. Timber leases on state lands are state-market.

There is some gradation within those four options, but they provide conceptual clarity. I will examine two competing models of drone airspace regulation, federal-administrative and state-market. As explained later in "Historical Administrative Assignment of Airspace and Routes," high-altitude airspace is assigned administratively (though much of that work simply codifies what the economically and politically powerful operators have reached consensus on through international standards bodies and trade associations). In the state-market model, in contrast, states (and their governmental subdivisions) play a major role in low-altitude drone operations as sovereigns. Regulation and airspace rights in this model are demarcated by state officials and leased via competitive bidding or sales to operators.

These models are intended as ideal types. Drone regulation in 10 or 20 years will be somewhere on a spectrum that includes them. Passenger drones, for instance, would likely benefit from a federal-market model similar to the radio spectrum.[20] Further, some state aviation officials would prefer something resembling a state-administrative model, though any state-centric model is likely to face federal agency and drone industry resistance. States have significant autonomy in a state-market model but, to be safe and practical, must share authority and collaborate with federal aviation regulators for the purpose of protecting the safety and operations of traditional aviation and interstate commerce.

For the reasons explained later in "Airspace as Property and State Powers in Airspace Policy," states and localities are likely to play a larger role in drone management than they do in traditional aviation regulation. Most surface airspace is private property, and state law delimits the contours of property rights. Perhaps more contentiously, administrative allocation of airspace seems ill-suited for the drone industry. As discussed later in the two sections "Historical Administrative Assignment of Airspace and Routes" and "Technology Lock-In," administrative allocation of airspace and terminals has already led to technology lock-in and anti-competitive route-squatting. In the alternative state-market model, market disposition and leasing of drone airspace offer many competitive, innovation, and practical benefits to administrative allocation.

AIRSPACE AS PROPERTY AND STATE POWERS IN AIRSPACE POLICY

Airspace is a valuable but underutilized natural resource. Like petroleum prior to 1850 or radio spectrum prior to 1900, airspace was virtually unused and valueless until inventors created technology to use and exploit it. The problem for those who favor a uniform federal law of drone airspace is that "airspace as property" has a long pedigree in American law. Absent a revolution in property rights jurisprudence, states, cities, and landowners will shape national drone management policy.

As an initial matter, a legal and policy analysis must separate airspace into two zones: surface airspace and nonsurface airspace. The boundary between the two has never been formally demarcated; nevertheless, courts, Congress, and state governments recognize and treat these zones differently.

Drone flights through or near surface airspace involve a mix of federal and state prerogatives, which are sometimes at odds. Congress hasn't brought clarity to the federal-state divide over drone airspace issues; and while Congress stands by, states are increasingly asserting their powers over drones and the use of surface airspace. Further, influential law drafters such as the Uniform Law Commission, American Bar Association, and American Law Institute are drafting airspace trespass provisions that provoke resistance from the drone operators and national regulators who favor unitary federal regulation.[21]

Nonsurface airspace is heavily used for interstate and international travel and thus is regulated almost solely by the federal government, a federal-administrative model of public resource regulation. Low-altitude airspace, however, has been tied closely to the surface land beneath it despite political efforts to separate land from airspace for more than a century. Surface airspace raises property rights, takings, and state powers issues that are negligible or absent for nonsurface airspace. The inseparable nature of land and airspace means that prospective drone regulation requires analysis of states' property laws and limits on federal and state governments' powers over airspace management.

In the late 19th and early 20th centuries, most Anglo-American courts and property theorists rejected the notion that "land" (thus trespass liability) projected infinitely upward.[22] The surface airspace, however, has long been treated as real property.[23] Anglo-American legal treatises from the 1840s onward noted that property could be partitioned horizontally[24] and that airspace—the "upper chamber" of a parcel of real estate—could be owned separately from the surface property.[25] As one early treatise noted, "It follows from this [*ad coelum* principle, the legal aphorism that land ownership extended "to the heavens"] that land may be divided horizontally as well as vertically, and the owner of land may divide and sell the space above the surface . . . as well as he can divide the surface into city lots."[26]

High-rise construction innovations at the turn of the 20th century increased the value of surface airspace as buildings could now occupy airspace high above urban land. New York's 1916 zoning law was the first to limit building size by volume—height and setback rules—and this accelerated the propertization of airspace in cities.[27] Airspace sales and transfers began in earnest in the 1920s,[28] especially after the Merchandise Mart development above Union Station in Chicago.[29] The attorneys negotiating the megaproject created and conveyed the world's first commercial "air lots"—three-dimensional, platted volumes of land—for the construction of department stores above the Chicago railroad terminal.[30] Air lots represented a new type of realty that the attorneys derived from common law principles of land partition.[31]

The proliferation of airspace property laws led the American Bar Association's president in 1930 to remark on the "increasingly common" practice of "leasing or selling air space."[32] By the 1960s, condominium laws simplified the process of demarcating fee simple interests in land in a vertical column,[33] and the creation and sale of airspace tracts separate from the land was routine.[34] Today, airspace and airspace lots are typically treated as a form of real property for tax and recording purposes.[35]

Owing to this principle that surface airspace is private property, under current constitutional understandings, federal regulation over surface airspace is limited by state sovereignty and takings considerations. Long before drones, these property principles crept into early aviation policy in the form of state claims of sovereignty to surface airspace. This low-altitude airspace was viewed as part of the underlying land, over which state governments had sovereignty.[36] Beginning in the 1920s, after the drafting of the Uniform State Law for Aeronautics, U.S. states began codifying their claims of sovereignty over surface airspace against the federal government.[37] At least 19 states have these laws today.[38]

Congress passed the Air Commerce Act in 1926 to bring some order to the regulation of the growing interstate and foreign air service industries. Included in that act was a declaration of "complete sovereignty of the airspace over the lands and waters of the United States."[39] Read in isolation, this provision is sometimes misinterpreted as a nationalization of airspace against state and local powers.[40] That interpretation is fairly easy to dismiss. The idea that this provision amounts to nationalization or preemption against the states was repudiated by the law's drafters,[41] the Senate legislative counsel,[42] and the Supreme Court.[43]

Further, the Supreme Court and lower courts have rejected federal regulators' authority to allow low-altitude flights without landowner permission or compensation. *United States v. Causby,* which involved the U.S. military flying planes that invaded the surface airspace over a farmer's property, has shaped airspace law and policy since it was decided in 1946. Surface airspace, the Supreme Court held, is part of land ownership. In particular, the Court said, overflights of government planes represent a permanent, physical invasion—a taking—if "they are so low and so frequent as to be a direct and immediate interference with the enjoyment and use of the land."[44]

In litigating *United States v. Causby*, the federal government argued that (a) flights at low altitude, if within "navigable airspace," cannot amount to a taking;[45] and (b) landowners do not own surface airspace—the "superadjacent airspace"—except that surface airspace occupied by buildings.[46] The Supreme Court rejected both arguments.

In rejecting the government's first argument, the Court held that when "navigable airspace" overlaps with the airspace that the landowner can "use in connection with the land," a taking can occur.[47] In rejecting the second argument, the Court held that landowners do own surface airspace above their land: "The landowner owns at least as much of the space above the ground as he *can occupy or use* in connection with the land."[48]

Also notable for our purposes, the Court drew on the earlier legal understandings and acknowledged and cited favorably North Carolina's claim to sovereignty to surface airspace in its takings analysis.[49] The Supreme Court reiterated in *United States v. Causby* that "while the meaning of 'property' as used in the Fifth Amendment [is] a federal question, 'it will normally obtain its content by reference to local law.'"[50] The Court has reiterated this reliance on state property laws in takings litigation in recent decisions.[51]

In 1958, Congress clawed back some regulatory authority over airspace from the states as jet travel created new airspace congestion (and collision risk) with the expansion of domestic airline routes nationwide. Nevertheless, as the Court reiterated in *Griggs*, another low-flying airplane case a few years after passage of the 1958 Federal Aviation Act, flights in navigable airspace, if disruptive and at a low altitude, can constitute a taking.[52] In the decades since, many courts have adopted an informal rule that flights below 500 feet get more scrutiny and are presumptive evidence of a trespass of private property.[53]

HISTORICAL ADMINISTRATIVE ASSIGNMENT OF AIRSPACE AND ROUTES

States and localities have played a small role in airspace management and disposition. States are capable of disposition of natural resources—such as timber lands, grazing lands, and water rights—but manned aircraft, like airliners, general aviation flights, and helicopters, transit mostly in nonsurface airspace, and most passengers and freight are part of interstate or international commerce. As a result, the federal government has played a dominant role in traditional airspace design and assignment. After a short period of market disposition—open bidding—on air routes in the 1920s, federal regulators aggressively stepped in and assigned airspace, routes, and scarce terminals, typically via administrative processes. The perils of that approach, particularly the anti-competitive behavior and the technology lock-in, are difficult to remedy once in place.

In the mid-1920s, Congress turned over airmail routes to private, startup airlines that bid on routes offered by the government. There was significant consolidation among these airlines, but new firms kept emerging and bidding on point-to-point mail routes.[54] Some leaders within industry started lobbying Congress for restriction mechanisms to limit competition for routes and strengthen airline finances, and Congress responded with the passage of the Air Mail Act of 1928. The act, however, did not stimulate the consolidation the industry had hoped for. In 1929, around 40 airline companies still had routes spanning the United States,[55] and they lobbied for more restrictions on entry.[56]

President Herbert Hoover appointed a new postmaster general, Walter Brown, in 1929. Brown believed that the airmail routes were illogical and that the bidding for routes encouraged cost-cutting that made planes unsafe and prevented the passenger airline industry from maturing.[57] Soon after his appointment, Postmaster General Brown drafted legislation in consultation with representatives from the major airmail and nascent passenger airline companies.[58] The proposals moved air routes out of the marketplace and into government and incumbent industry control. Brown's draft legislation, for instance, gave the postmaster general broad authority to give no-bid route contracts to airlines "in the public interest."[59] That discretion to provide no-bid contracts faced opposition in Congress and was stripped out, but much of the draft legislation granting discretionary powers to the postmaster general was passed by Congress in the McNary-Watres Act of 1930.[60] Among other things, the McNary-Watres Act provided higher postal compensation to airmail airlines that had larger, passenger-grade aircraft.[61] Bidding on new routes was restricted to firms that had experience on a 250-mile route, a restriction intended to shut out new entrants.[62] While the provision allowing Brown to grant no-bid route contracts was omitted, the law still authorized the postmaster general to *extend* existing routes without a competitive bidding process.[63]

A later investigation by the Franklin Roosevelt administration revealed the unseemly market division Brown orchestrated. Investigators found that within weeks of the McNary-Watres Act's passage Postmaster General Brown had summoned the larger airlines to a closed-door meeting and instructed them to consolidate their airlines and divide the new transcontinental routes he had designed among themselves.[64] By law, Brown had to advertise and accept bids for the routes, but under pressure from Brown, the airlines privately agreed not to bid on any routes except those the postmaster general had informally designated for them in those 1930 meetings.[65]

However, in the weeks of industry negotiation that followed, the airlines failed to come to an agreement satisfactory to Brown, and he allocated routes among them.[66] Three airline operators emerged from the Brown meetings with 90 percent of the nation's airways.[67] United Aircraft & Transport was assigned the northern route, Transcontinental & Western Air (TWA) was assigned the center route, and American Airways was given the southern route. The Big Four were established when Eastern Air Transport accumulated the routes along the Eastern Seaboard.

This gift of routes to a few airline operators was revoked briefly by the Roosevelt administration, but President Roosevelt merely initiated another government debacle in aviation. In part because of the anti-competitive division and assignment of routes by Postmaster General Brown and the airlines, the Roosevelt administration canceled all existing airmail contracts via executive order in early 1934.[68] In February 1934, the administration nationalized the airmail system, transferring airmail duties to the Army Air Corps; but nationalization was an unmitigated, albeit brief, disaster. In the first 10 weeks, the ill-trained, ill-equipped Army pilots were involved in 66 accidents and 12 pilot deaths.[69] By April, after public outcry, the Roosevelt administration had no choice but to reprivatize airmail.[70] Under the ensuing Air Mail Act of 1934, the three participating companies from the Brown meetings were barred from the bidding process, but the dominant airlines were largely reconstituted with some changes in corporate form, leadership, and names.[71] At the behest of carriers, the Civil Aeronautics Board was formed by the Civil Aeronautics Act of 1938 to serve as a third-party governance structure to restrict competition and set rates.[72]

This administrative partitioning and assignment of national airways—almost entirely the designs of a sole postmaster general determined to consolidate the airline industry and give no-bid contracts in defiance of Congress and reticence from industry—has had long-lasting anti-competitive effects. The administrative assignment of airspace, begun in 1930, led to large economic distortions that are occasionally still visible today.

In particular, congestion at airports is a symptom of administrative assignment of routes and terminals. Airport congestion was light in the early years but proved unworkable once aviation matured, particularly after the spread of jet engines in the 1960s. While commercial airlines had fixed routes, the remaining airspace was essentially a regulated commons—smaller planes could freely use airspace and terminals on a noninterference basis with the airlines. Airspace and terminals in urban areas had too many users, but authorities' attempts to rationalize airspace use with pricing were thwarted.

> For example, in 1968, nearly one-third of peak-time New York City air traffic—the busiest region in the US—was general aviation (that is, small, personal) aircraft. To combat severe congestion, local authorities raised minimum landing fees by a mere $20 (1968 dollars) on sub 25-seat aircraft. General aviation traffic at peak times immediately fell over 30%—suggesting that a massive amount of pre-July 1968 air traffic in the region was low-value. The share of aircraft delayed by 30 or more minutes fell by half, from 17% of flights to about 8%.[73]

This pricing of airspace and airport access—providing a peek at the beneficial power of pricing scarce resources—did not last long. The aviation industry and its advocates, accustomed to and adapted to administrative disposition of airspace and terminals, resisted pricing and market reforms.

Regulators fell back on a new type of rationing via the creation of *airport slots*: designated time periods, say, 15 or 30 minutes in length, in which a plane may take off or land at an airport. The slots at major hubs at peak times of day are extremely scarce because demand is so great. Slot rationing began in the United States as a temporary program in response to an immediate crisis. Yet today that temporary program is still in place—slots are currently used to ration access at LaGuardia, John F. Kennedy International, and Ronald Reagan Washington National airports. And while it doesn't use formal slot rationing, the FAA also administratively rations access at four other busy hubs: Chicago O'Hare International, Newark Liberty International, Los Angeles International, and San Francisco International airports.

Market reforms, proposed since at least the 1960s, have gotten little traction. In 2008, at the tail end of the George W. Bush administration, the FAA proposed to auction some slots in New York City's three airports. The plan was delayed by litigation from incumbent airlines and an adverse finding from the GAO. With a change in administration, the FAA under President Barack Obama rescinded the plan in 2009.

Before that rescission, the question of the propriety of giving valuable slots to airline operators was hinted at in GAO's criticism of the slot auction plan in 2008:

> FAA's argument that slots are property proves too much—it suggests that the agency has been improperly giving away potentially millions of dollars of federal property, for no compensation, since it created the slot system in 1968.[74]

Then, suddenly, federal policy reversed on the question of auctioning slots, and the FAA auctioned two dozen high-value slots in 2011. Delta and US Airways wanted to swap some 160 slots at New York and Washington, DC, airports. As a condition of the mega-swap, the FAA required they divest 24 slots at those popular airports, which the agency auctioned to new entrants. Seven low-fare airlines bid in the auction, and JetBlue and WestJet won the divested slots, paying about $90 million combined.[75]

The (short-lived) landing fees in the 1960s and slot auctions in 2011 reveal what is obvious to transportation economists—the long-standing policy of administrative assignment of terminal access and airspace access conceals their value and distorts their use. Reliance interests by industry are significant, however, and today airspace access and terminal access are typically negotiated by operators at forums provided by the FAA and by international industry associations. The legacy administrative processes allow incumbent airlines to dominate the slot allocations via international conferences and national regulations that require "grandfather" slot usage. In a recent article, the *Wall Street Journal* reported that airlines are reluctantly ceding more power to airports in the assignment of slots.[76] This is a signal in the long-running tug-of-war between airports and airlines. Airports generally want to open slots for new competitors, while incumbent airlines do not.

Airspace and terminal access clearly are highly valued. (Congress has not yet asked GAO to follow up on its provocative question in its 2008 analysis and assess the value of slots disposed of using the federal-administrative allocation method over decades.) However, attempts at market reforms are fiercely resisted by industry and within government. The reforms have been incremental and minor to date. The government favored certain aviation operators and gifted routes and terminals during the industry's infancy, policies that have proven very difficult to dislodge.

TECHNOLOGY LOCK-IN

An understudied side effect of administratively rationing airspace and terminal access is that regulators needed an interconnected, national air traffic control system to prevent unsafe congestion at high-traffic airports. Shared terminals and airspace require manual or automated deconfliction between users. Updating the air traffic control system has proven very costly and time consuming. Many of the airspace users and constituencies—airlines, general aviation flyers, charter planes, information technology system suppliers, air traffic controllers, aircraft manufacturers, among others—have adapted to current policies and legacy technology and strongly resist updating the interconnected system.

The rigidity of the system is well known to the industry and regulators, but the failure to upgrade has uncaptured benefits that can only be speculated about. Transportation economist Clifford Winston has written about the difficulties in upgrading the current air management system.[77] In the early 1980s, the FAA announced plans to develop an advanced automated system that was scheduled to be completed by 1991 at a cost of $12 billion. Writing in 2014, Winston noted that the proposed upgraded system was more than two decades late, billions of dollars over budget, and more modest than planned.[78] The automated system is still a long way off, and the FAA has turned its attention to a modest transitioning of the current radar-based system to a satellite-based system.[79]

John Palfrey and Urs Gasser, law scholars at Harvard University's Berkman Klein Center, have studied interconnected networks and note, "This problem of lock-in is one of the core puzzles of interoperability. . . ."[80] For

Palfrey and Gasser, air traffic control stands out for its interconnection drawbacks; they note how difficult it is to integrate "plainly superior technologies" like the GPS into air traffic management.[81] The difficulty of improving traditional air traffic management stems from "the deeply rooted interoperability of the current system."[82] Palfrey and Gasser's conclusion has sobering implications: "[I]t is very hard to envision what a successful interoperability strategy for the next generation of air traffic control systems will or should look like, because there are so many stakeholders around the world and so many different technologies involved."[83]

As a result, U.S. airspace management is still very interconnected and manually operated, and it protects the reliance interests of incumbent users. One illustration of this fact is that the U.S. airspace command center "has seats on the floor for industry organizations to take part in the decision-making process."[84] While safe and reliable, the existing system requires tremendous regulatory overhead and cost—more than 16,000 FAA employees and contractors work on upgrading the current system—and it is very difficult to update with new entrants (who would disrupt the stability of the system) or new technology (because piecemeal updates can break an interconnected system).[85]

THE STATE-MARKET MODEL APPLIED TO DRONE AIRSPACE

Federal and state policymakers should avoid extending traditional aviation's federal-administrative framework to drones. Instead, lawmakers should consider a framework of cooperative federalism to quickly integrate drones into U.S. airspace while avoiding controversy and litigation between the federal and state governments. Cooperative federalism tends to work when the state leaders have similar interests and the issue is nonpartisan.[86] Federal and state aviation officials and lawmakers, regardless of party, seem generally amenable to widespread commercial drone use. Drones are used in a wide variety of industries—agriculture, public safety, medicine, logistics, utilities, photography, among others—and offer new education tracks and skilled jobs.

To support that case, I will apply the stricter state-market model to drone airspace regulation. The state-market model avoids the legal and practical problems associated with the federal-administrative model. In short, using this model, state authorities would demarcate aerial corridors above public roads and auction or lease those volumes of airspace to drone operators. Clarifying state and private landowners' authority over surface airspace avoids litigation over federal preemption. Further, the demarcation and leasing of surface airspace—in particular, aerial corridors for drone operations—allows the public resource to go to its highest-valued uses and gives new firms an opportunity to enter the marketplace.

The jurisdictional problem arises because surface airspace *is land*—subject to state police powers—and *is navigable airspace*—subject to federal regulation. As mentioned earlier (see the section Airspace as Property and State Powers in Airspace Policy), many states have claimed sovereignty over low-altitude airspace. Not only do states claim sovereignty to surface airspace, the foundation for a state-market system in airspace is already in place: more than 20 states expressly allow state and municipal officials to lease airspace above public land or public easements, such as the airspace above roadways.[87]

Airspace leasing is not new. This practice of roadway airspace leasing was formalized in a 1961 amendment to federal highway laws allowing states and cities "to use or permit the use of the airspace above and below . . . the highway pavement for such purposes as will not impair the full use and safety of the highway."[88] State departments of transportation (DOTs) began leasing public airspace—typically above the right-of-way—in earnest in the 1970s and 1980s as a revenue source.[89] Since 1986, the Federal Highway Administration has encouraged airspace leasing and offered technical assistance to state DOTs to routinize the practice.[90]

To date, this authority has not been used for drone highway use, but drone use, within railroad and utility rights-of-way and aerial corridors, has occurred. A national policy of allowing airspace leasing to drone companies would have an immediate and beneficial effect on drone services. Possibly millions of miles of aerial corridors would be made available. Roadways and their accompanying airspace represent a huge amount of unused, non-revenue-generating real estate. Most major cities' road systems take up 25–35 percent of the city's land area.[91] For more suburban areas the percentage is smaller, around 15–20 percent.[92] According to estimates using Federal Highway Administration data, "the amount of existing [right-of-way] that is a part of the National Highway System [is] 3,000–6,000 square miles," which is about the size of Connecticut.[93] Further, there are "more than 8 million lane miles of public roadways under state DOT supervision."[94] This existing state authority over airspace leasing complements drone technology, particularly in urban areas, because drones will likely depend on locally controlled infrastructure, including roadside beacons, GPS repeaters, and 5G antennas.[95]

Demarcating drone highways above roadways also avoids predictable trespass, nuisance, and taking litigation from landowners. The downside of private operation of airports is that, historically, private airports are more likely to lose trespass and nuisance lawsuits than public airports.[96] In contrast to publicly operated airports, privately owned airports face court injunctions.[97] Court penalties also seem to be stiffer, including daily damages for continuing operations.[98]

Because drone lawsuits tend to be in state courts, drone, unmanned aircraft system (UAS) traffic management (UTM), and "droneport" operators are relatively vulnerable to lawsuits. Much more than federal courts, which tend to limit inverse condemnation to overhead flights, state courts interpret state constitutions' conception of inverse condemnation to include aircraft noise over adjacent properties.[99] As a result, if drone operations are not over roads—which use existing public rights of way—operators face potential lawsuits from virtually any resident subject to overflights because most drones are near the surface the entire flight, not simply during takeoff and landing (as in traditional aviation). With drone overflights, landowners face not only nuisance and trespass but, over years, the prospect that drone operators could acquire a prescriptive easement to the landowners' airspace, potentially entitling drone operators to enter property to cut trees or prevent new construction on the land.[100]

In the real world, a compromise on purely state authority would necessarily need to be overlaid with significant federal input and cooperative federalism. The federal role would be to demarcate drone airspace covering geographic regions that pose a *de minimis* risk to traditional, manned aviation.[101] The FAA and other U.S. DOT agencies would focus on aerial rights above local roads near airports and heliports and above interstate highways, where federal interests are greatest. They would also expand their existing air rights leasing expertise to states needing it. Finally, under a cooperative federalism framework, the federal government would prohibit rules and policies that prohibit or have the effect of prohibiting interstate drone commerce.

Perhaps the closest cooperative federalism model and analog is telecommunications. The construction and operation of droneports and drone highways, like telecommunications, will require local zoning permits and private property. In telecommunications, while the Federal Communications Commission (FCC) has sole authority over communications devices and interstate communications, the agency does not have the power to pick and choose where all telecommunications facilities are installed. The construction of cell sites and conduit is governed by state and local police powers. Despite those powers, Congress authorizes the FCC to preempt state or local rules that "may prohibit or have the effect of prohibiting the ability of any entity to provide any interstate or intrastate telecommunications service."[102] There should be a similar preemption provision for state drone regulation.

THE FEDERAL-ADMINISTRATIVE MODEL APPLIED TO DRONE AIRSPACE

The state-market model will face opposition for its state-centric focus of authority and for its market disposition of a public resource. The primary alternative model for drone regulation is the federal-administrative model, which current drone policy resembles. Under the traditional airspace model, routes, slots, and terminal locations are assigned via federal regulation, with significant input from legacy operators and industry insiders.

In summer 2020, the FAA came out in favor of "urban air mobility corridors"—point-to-point aerial highways that new electron vertical takeoff and landing aircraft, helicopters, and passenger drones will use.[103] While the endorsement of aerial corridors for these new services is commendable,[104] the FAA's plan for allocating and sharing those corridors is largely to let the industry negotiate it among themselves.

While federal policy is less developed when it comes to small drones, a corridor system is also developing. This policy trajectory won't end well; it resembles a gentler, more formal version of the postmaster general letting the nascent airlines in the 1930s divvy up air routes. The result in traditional aviation is that decades later the FAA is still refereeing industry fights over routes and airport access. Lacking a meaningful price mechanism, the politically powerful have a significant advantage.

The assignment of drone airspace will not look exactly like the anticompetitive route assignments in 1930. It is recognized, however, that there is a race to claim routes and that drone operators would benefit from route squatting. As McKinsey analysts noted in 2019 about urban air mobility, "first movers will have an advantage by securing the most attractive sites along high-traffic routes."[105]

A similar "race to the regulator" is already underway to get UTM systems and drone services approved.[106] Administrative assignment benefits those with some combination of promising (but unproven) technology, legal acumen, and political savvy. In short, the well-connected will gain airspace access and approval to offer drone management services. This technology is in its infancy,[107] and there is no promise that political insiders have the best technology or best operations, or that they value the airspace access most. Further, this traditional model entrenches slow-moving incumbents. It will be difficult if not impossible for later competitors to dislodge them from high-revenue routes and droneports.

Regulators and lawmakers have a choice to make: Which airspace regulatory model on the federal-state and administrative-market axes will minimize costly litigation, ensure aviation safety, and encourage rapid deployment of new drone services? There is significant inertia behind the federal-administrative model that is used in traditional aviation, but small drones raise many more property rights and federalism issues. To prevent the technology lock-in and undue first-mover advantages seen in traditional aviation—which are damaging for an innovative sector—lawmakers and industry should immediately consider the feasibility of airspace auctions and leases.

CHAPTER FIVE

WHO OWNS THE SKIES? *AD COELUM*, PROPERTY RIGHTS, AND STATE SOVEREIGNTY

LAURA K. DONOHUE

Cuius est solum, eius est usque ad coelum et ad inferos.
Translation: *Whoever owns land it is theirs up to the heavens and down to hell.*
—Franciscus Accursius

In *Glossa Ordinaria* (1220–1250 CE), a seminal work that rendered earlier scholarship obsolete and for 500 years served as the authoritative statement of Justinian law, Franciscus Accursius established that whoever owned the land controlled everything from the heavens above to hell below.[1] It was far from a novel concept. For more than a millennium, the concept *ad coelum* had distinguished between the skies aloft and terrestrial matters. In his 54 BCE epic poem embracing Epicurean philosophy, *De Rerum Natura*, Titus Lucretius Carus equated *ad coelum* with the sky, currents of air above the earth, and the heavens themselves.[2] Virgil's *Aeneid*, written in 19 BCE, treated *ad coelum* in similar fashion.[3] *De caelo servare* came to mean to observe lightning (as an omen) and *de caelo tactus* to be struck by lightning, even as *toto caelo* came to mean something like "utterly"—that is, to the full extent of the heavens.[4] A simple principle applied: land ownership conveyed control of three-dimensional space.

The Justinian concept worked its way into English common law and thence to the colonies, for incorporation into Anglo-American jurisprudence. It was only in the 20th century, with the advent of air travel, that limits in the form of effective possession began to appear. An uneasy compromise of 500 feet was proposed, with property owners' rights below that level secured and subject only to state and not federal domain.

In light of the history of the doctrine of *ad coelum*, as well as the states' preeminent role (secured by the Tenth Amendment) in regulating property and airspace up to the 500-foot level, it is remarkable that the federal government has begun to claim that it controls everything above the blades of grass. In an article aimed at "busting myths about the FAA," the Federal Aviation Administration (FAA) asserted, "The FAA is responsible for the safety of U.S. airspace from the ground up."[5] Jim Williams, the head of the FAA's unmanned aircraft systems (UAS) Integration Office, baldly asserted, "If you are flying in the national airspace system, FAA regulations apply to you. The definition of the national airspace system is anywhere where aircraft can safely navigate. So by definition then, these quadcopters are what have extended the national airspace down to the ground."[6] In a response to a motion to dismiss on a case involving a drone operating at altitudes of less than 400 feet above the University of Virginia, the FAA administrator similarly asserted, "the FAA's mandate to regulate the use of all airspace necessary to 'ensure the safety' of aircraft, for 'protecting and identifying' those aircraft and for 'protecting individuals on the ground' is not confined solely to the 'navigable airspace.'"[7] This chapter challenges those statements, demonstrating that history and law establish that property owners, and the states, control the airspace adjacent to the land.

AD COELUM IN ENGLISH COMMON LAW

King Edward I's invitation to Accursius's son, Franciscus, to come to England paved the way for the incorporation of *ad coelum* into English common law. In 1274, Franciscus took up the post at Oxford, lecturing on Roman Law.[8] Courts accredit that moment as the point at which the term entered the legal lexicon.[9] English treatise writers went on to acknowledge the importance of *ad coleum*. In his *First Institute of the Laws of England*, the prominent jurist Sir Edward Coke observed in relation to *terra*, "And lastly the hearth hath in law a great extent upward, not only of water as hath been said, but of aire, and all other things even up to heaven, for *cujus est solum ejus est usque ad coelum*, as it is holden."[10] He cited two cases related to a dispute between a landlord and tenant challenging who, under the lease in question, owned six young goshawks.[11] A third case centered on the rights held by the Bishop of London to herons and shovelers nesting in trees that he had leased.[12] The court reasoned that ownership of the land includes the roots of the trees—and the branches in the airspace above. In each situation, the owner maintained an interest in the full use and enjoyment of the property.[13]

At the time that Coke wrote, there was a sharp distinction in the law between the right in land (i.e., arising out of the ownership of the land), and the right of peaceful enjoyment of property. For the former, certain rights attained to the dignity of ownership or possession. Others—such as piscary and turbary, rights of ingress and egress, ancient rights of firebote and fencebote, and the right of escheat (which belonged to the Crown)—did not.[14] Violating the airspace above the land constituted an interference in the enjoyment of the property.

Two cases reported by Coke followed his exposition in the *Institutes.* In *Penruddock's Case* a writ of *quod permittat prosternere*, commanding that the defendant allow the plaintiff to abate a nuisance or appear in court and show cause why not, attached.[15] Under the writ, plaintiffs were entitled to a judgment for abatement and corresponding damages. In this case, the court determined that rain draining from an overhanging building adjacent to the plaintiff's land was a nuisance.[16] In 1611, the court again applied the ancient writ of *quod permittat* in *Baten's Case* to find an overhang to be a nuisance and ordered its abatement.[17]

For centuries, the maxim *ad coelum* stood, recognizing property owners' rights to the airspace above their land. According to the facts of the case, courts alternately found either that a trespass had occurred or that a nuisance resulted from incursions into the space above the property. In 1753, Sir William Blackstone delivered a series of lectures at Oxford, published in four volumes in 1765–1769. In volume II of the

Commentaries on the Law of England, Blackstone explained the importance of *ad coelum* for understanding the rights conveyed in real property:

> Land hath also, in its legal signification, an indefinite extent, upwards as well as downwards. *Cuius est solum, eius est usque ad coelum*, is the maxim of the law, upwards; therefore no man may erect any building, or the like, to overhang another's land: and, downwards, whatever is in a direct line between the surface of any land, and the center of the earth, belongs to the owner of the surface.[18]

The control of the land and the air above it meant that adjacent landowners were restricted in what they could do above their neighbor's property as well as below it. There was no limited right of access conveyed by the neighbor's interests. This was, for Blackstone, aptly illustrated in mining areas where such rights were well recognized. He continued, "the word 'land' includes not only the face of the earth, but everything under it, or over it."[19] So when *land* was conveyed, everything else—air, water, houses, and mines below the surface—transferred.[20]

The 17th- and 18th-century cases dealt largely with matters related to hunting, things falling on property from above or outside the boundaries, and stationary structures overhanging adjacent land. But in 1783 an entirely new category emerged: Joseph-Michel and Jacques-Étienne Montgolfier invented the *globe aérostatique*, raising questions associated with overflight. An 1815 case presaged the legal questions that would later arise: *Pickering v. Rudd* addressed a landowner's decision to cut back his neighbor's Virginia creeper that had grown onto his house and to then hang a board that projected across the property line. Lord Ellenborough in *dicta* infamously posed a *reductio ad absurdum*: if the board be a trespass, might an aeronaut be "liable to an action of *trespass quare clausum fregit* at the suit of the occupier of every field over which his balloon passes in the course of his voyage"?[21] He compared firing a bullet over land to circumvention of a balloon, proposing that the latter might constitute a nuisance, but not a trespass.

English courts flatly rejected Ellenborough's reasoning and found interference in the airspace above land to be either a nuisance or a trespass (and sometimes both), upholding the rule from antiquity that protected property owners in the use and enjoyment of their property. In 1845, for instance, in *Fay v. Prentice*, a cornice jutting out over a neighboring garden and funneling rainwater onto the adjacent land constituted a nuisance.[22] Judge Thomas Coltman, citing *Baten's Case*, determined that it did not matter whether water had actually caused damage. The mere overhang was sufficient. By it, the plaintiff "had been greatly annoyed and incommoded in the use, possession, and enjoyment of his messuage, garden-ground, &c."[23] Judges William Henry Maule and Cresswell Cresswell concurred.[24] In 1855, the Court of the Exchequer of England determined that wires above property occupied the land beneath.[25] The judge, Sir Charles Edward Pollock, explained, "Land extends upwards as well as downwards, and whether the wires and posts are fixed above or below the surface, they occupy a portion of the land."[26]

Case after case followed course. In the 1865 case of *Kenyon v. Hart*, a hunter shot a cock pheasant that was flying over his neighbor's property and then went to retrieve it.[27] Justice Colin Blackburn took the occasion to remark on Ellenborough's reasoning, "I understand the good sense of that doubt but not the legal reason of it."[28] Five years later, in *Corbett v. Hill*, the court considered a conveyance on sale that established an underground flying freehold—that is, a section of freehold property extending below ground.[29] The plaintiff, who owned two houses next door to each other, sold the adjoining property, which supported one of the rooms of the house that he retained. Sir William James Montgomery-Cuninghame determined that the buyer (the owner of the second property) had a property right in the air above his land, making any intrusion in the space a trespass.[30] That same year, three of four judges ruled an equine owner negligent when his horse kicked through a fence and thereby injured a horse on a neighbor's property.[31] The justices (one citing *ad coelum*) agreed that by entering the airspace above the ground, the horse had trespassed on the adjoining land.

In 1880, an action was brought to restrain the 12th Middlesex Volunteer Corps from shooting guns on Wimbledon Common to the detriment of an adjacent landowner's property.[32] According to the suit, "splashes of bullets and flattened bullets fell on the plaintiff's land, so as to substantially affect the ordinary use and enjoyment of the property."[33]

The defendants in *Clifton v. Bury* claimed that the Putney and Wimbledon Commons Act, 1871, reserved the land for public use, that even on the admitted facts no trespass had actually occurred, and that only those who actually fired the weapons might be held responsible.

Justice Henry Hawkins, ruling in favor of the plaintiff, noted that the statute in question did nothing to divest property owners of the rights they held in land or the air adjacent: "Nobody," he concluded, "could suggest seriously that the line of fire over Newlands Farm formed a part of the common."[34] The statute had explicitly *not* deprived landowners of any of their property rights in the fields adjacent to the commons, which meant that they continued to have a right to the airspace. Every bullet that crossed the land and fell, moreover, "materially interfere[d] with the plaintiff's ordinary use and enjoyment of his farm," constituting "a series of trespasses of an actionable character."[35] Although no bullets had actually been proven to have crossed the land, and while no injury had been occasioned to the plaintiff, the use of the airspace had "rendered the occupation of that part of the farm less enjoyable than the plaintiff was entitled to have it."[36]

Using similar logic, the House of Lords in *Lemmon v. Webb* determined that a landowner was free to cut off his neighbor's tree as it overhung his property, without first obtaining permission or giving notice of his intent to do so.[37] The court determined that allowing boughs and roots to extend onto adjacent property was a nuisance and that any person whose land was so affected had the right to abate it.

The law of torts thus consistently recognized the importance of *ad coelum*. Neither entry onto another's land (trespass *quare clausum fregit*) nor taking someone's goods (trespass *de bonis asportatis*) required the use of force, the breaking of an enclosure, the transgression of a visible boundary, or even unlawful intent.[38] Nor did actual damage have to occur. In his authoritative treatise first published in 1887, Sir Frederick Pollock recognized the ancient protections extended by common law to property rights. He quoted *Entick v. Carrington*, "Every invasion of private property, be it ever so minute, is a trespass."[39] "There is no doubt," Pollock wrote, "that if one walks across a stubble field without lawful authority or the occupier's leave, one is technically a trespasser."[40] Even "[l]oitering on a highway . . . for the purpose of annoying the owner of the soil in his lawful use of the adjacent land, or prying into his occupations there, may be a trespass against that owner."[41]

Pollock spoke directly to *ad coelum*: "It has been doubted whether it is a trespass to pass over the land without touching the soil, as one may in a balloon, or to cause a material object, as shot fired from a gun, to pass over it."[42] Pollock, like the English courts before him, roundly rejected Lord Ellenborough's ruminating in *dicta* in *Pickering*, stating, "Fifty years later Lord Blackburn inclined to think differently, and his opinion seems the better."[43] Pollock noted that "wrongful entry on land below the surface, as by mining" was "prominent in our modern books."[44] *Pari passu*, "[i]t does not seem possible by the principles of the common law to assign any reason why an entry *at any height* above the surface should not also be a trespass."[45] While it may be improbable for someone crossing over at a great height to cause any actual damage to the land below, the question of damage was utterly "irrelevant": if mere trespass was sufficient for the surface, so, too, for the air above. For Pollock, "it would be strange if we could object to shots being fired across our land only in the event of actual injury being caused."[46]

In later editions, Pollock specifically called out navigable aircraft:

> Clearly it would be a trespass to sail over another man's land in a balloon (much more in a controllable air-craft) at a level within the height of ordinary buildings, and it might be a nuisance to keep a balloon hovering over the land even at a greater height.[47]

Pollock acknowledged that "the projectiles of modern artillery, when fired for extreme range, attain in the course of their trajectory an altitude exceeding that of Mont Blanc or even Elbruz."[48] But while a projectile at such a height might not constitute a trespass, a different situation holds closer to earth.[49] There, undoubtedly, it is a trespass, and not a mere nuisance, that occurs.

Ad coelum applied regardless of whether land was held privately or for public use. In 1903, for instance, the question arose of whether a local highway authority could restrain the running of a power cable 34 feet above Regents Park Road in London.[50] The court noted that the council held the right to the property, which included, "All the stratum of air above the surface and all the stratum of soil below the surface which in any reasonable sense can be required for the purposes of the street, as street."[51]

THE AMERICAN CONTEXT

The American colonies, and later, states, adopted English common law. Professor James Kent's *Commentaries on American Law*, adapted from lectures that he presented at Columbia Law School starting in 1794, recognized its importance in early America:

> The common law so far as it is applicable to our situation and government has been recognised and adopted as one entire system by the Constitutions of Massachusetts, New York, New Jersey and Maryland. It has been assumed by the courts of justice or declared by statute, with the like modifications, as the law of the land, in every state. It was imported by our colonial ancestors, as far as it was applicable, and was sanctioned by royal charters and colonial statutes. It is also the established doctrine that English statutes passed before the emigration of our ancestors, and applicable to our situation, and in amendment of the law constitute a part of the common law of this country.[52]

It was not the entire body of law that transferred, but only those measures that applied to the colonists in their new conditions.[53] Upon independence, the new state legislatures and courts explicitly incorporated English common law through their state constitutions, as well as by statute.

In the Commonwealth of Massachusetts, the state constitution declared that all law previously in effect and practiced in a court of law would remain in force.[54] In Virginia, statutory provisions explicitly incorporated the common law.[55] Upon becoming a state in 1790, Vermont similarly specified, "So much of the common law of England as is applicable to the local situation and circumstances, and is not repugnant to the constitution or laws of this state, shall be deemed and considered law in this state, and all courts are to take notice thereof and govern themselves accordingly."[56]

This practice continued well into the 19th century, as new states were admitted to the union. Although California did not gain statehood until

1850, it still passed a general law stating, "The common law of England, so far as it is not repugnant to or inconsistent with the Constitution of the United States, or the Constitution or laws of the state of California, shall be the rule of decision in all the courts of this state."[57] Similar provisions marked the statute books of Illinois, Indiana, Missouri, Nebraska, Wisconsin, and other states.[58] Writing in *Lyman v. Bennet*, the Michigan Supreme Court explained:

> Practically the common law has prevailed here in ordinary matters since our government took possession, and the country has grown up under it. . . . A custom which is as old as the American settlements, and has been universally recognized by every department of government, has made it the law of the land if not made so otherwise. Our statutes, without this substratum, would not only fail to provide for the great mass of affairs, but would lack the means of safe construction.[59]

The court, as a result, was "of opinion that questions of property not clearly excepted from it must be determined by the common law."[60]

The concept of *ad coelum* was among the provisions of common law integrated into property law. Thus, in the 19th century, American state courts routinely ruled that overhanging branches constituted both a trespass and a nuisance.[61] Eaves extending over adjacent property were considered a trespass.[62] The same was true of cornices, windows, roofs, walls, or any part of any building: *any* incursion into the airspace of an adjacent landowner amounted to trespass on the property owner's land.[63] At other times, the courts considered the overhang to be a nuisance.[64] While there was some conflict over whether or not ejectment was a proper means of redress, trespass and nuisance both tended to result in an injunction as a means of redress.[65]

In the 1897 case, *Metropolitan West Side Elevated R.R. Co. v. Springer*, the construction of a railway across an alley amounted to a taking of the property by a landowner.[66] Illinois Supreme Court Chief Justice Jesse J. Phillips wrote, "It is a maxim of the law, *Cujus est solum ejus est usque ad coelum*, and by taking one foot on which the pillars were placed, on the south side of the alley, the rights reserved in the deed were invaded . . . by the projecting super-structure."[67]

The basic precept in dozens of state cases was that *ad coelum* granted landowners control above and below the property to which they held title. Thus, the decision to build a structure—or to remove it—lay within the property owner's domain.[68] The building was inseparable from the underlying land.[69] Landowners had a right to light in the air above it.[70] And they had a right to use of gravel, water, minerals, natural gas, petroleum, and other natural resources below it.[71] The same rights carried with the grant of an easement.[72]

Well into the 20th century, state courts continued to view such invasions of airspace as a trespass, a nuisance, or both.[73] It was not necessary for the item entering the airspace ever to touch the ground. Penetration of the column of air over the property was sufficient. Thus, a cornice, even when it didn't interfere with the plaintiff's use of the property, still interfered with the plaintiff's property rights, which extended upward indefinitely and which could be enforced against invasion to the same extent as surface rights.[74]

Even when the invasion is temporary and does no actual damage, it still violates the property owner's rights. In *Herrin v. Sutherland,* the Montana Supreme Court determined that shooting at duck flying over someone's property constituted trespass.[75] Chief Justice Llewellyn Callaway observed, "[I]t seems to be the consensus of the holdings of the courts in this country that air space, at least near the ground, is almost as inviolable as the soil itself."[76] There did not have to be any threat occasioned by the invasion of airspace to be considered a trespass: merely extending an arm over a neighbor's land to retrieve a ladder was considered sufficient.[77] In almost all of these cases, the courts looked to *ad coelum* to guide their thinking.[78]

The Supreme Court shared the states' approach, considering the firing of a gun across an individual's property to be a trespass. The routine discharge of a battery amounted to a taking.[79] In *Portsmouth Harbor Land & Hotel Company,* Justice Oliver Wendell Holmes Jr., writing for the Court, noted, "If the United States, with the admitted intent to fire across the claimants' land at will, should fire a single shot or put a fire control upon the land, it well might be that the taking of a right would be complete. But even when the intent thus to make use of the claimants' property is not admitted, while a single act may not be enough, a continuance of them in sufficient number and for a sufficient time may prove it. Every successive trespass adds to the force of the evidence."[80]

The decision was not unique to weapons. Any extended use of property owners' airspace by the government constituted takings. When telephone wires became standard for carrying communications, for instance, cases routinely recognized the importance of *ad coelum*. In the 1906 case of *Butler v. Frontier Telephone Company*, the New York Supreme Court explained:

> What is "real property"? What does the term include so far as the action of ejectment is concerned? The answer to these questions is found in the ancient principle of law: "*cujus est solum, ejus est usque ad coelum et ad inferos*." The surface of the ground is a guide, but not the full measure. "*Usque ad coelum*" is the upper boundary, and while this may not be taken too literally, there is no limitation within the bounds of any structure yet erected by man. So far as the case before us is concerned, the plaintiff as the owner of the soil owned upward to an indefinite extent. . . . According to fundamental principles and within the limitations mentioned, space above land is real estate the same as the land itself. The law regards the empty space as if it were a solid, inseparable from the soil, and protects it from hostile occupation accordingly.[81]

The court went on to underscore the point: "Unless the principle of "*usque ad coelum*" is abandoned, any physical, exclusive, and permanent occupation of space above land is an occupation of the land itself and a disseisin of the owner to the extent."[82] Land ownership extended downward as well. Thus, in 1878, the Pennsylvania Coal Company conveyed the surface of a plot of land but explicitly retained the right to mine underneath it.[83]

Efforts by the government to overregulate mineral extraction similarly reflected *ad coelum*, triggering protections against takings. In 1921, the Commonwealth of Pennsylvania enacted a statute preventing coal mining that could possibly affect the integrity of any surface land. The Supreme Court determined that while a state may pass laws in the valid exercise of its police powers that have an incidental impact on property values, when they causes sufficient diminution in property value, the state must take the land by eminent domain and provide compensation.[84]

So, too, was the construction of an underground sewer considered a taking—merely on the grounds that the property owners were deprived of *potentially* using the space occupied by the pipes.[85]

A similar approach to *ad coelum* held in regard to land held by public authorities. As explained in Professor Eugene McQuillin's *The Law of Municipal Corporations,*

> The public right to the use of streets goes to the full width of the street, and extends, indefinitely upward and downward. On the ground, therefore, of failure to exercise ordinary care to keep public ways in a reasonably safe condition for travel, municipal negligence may be established, on the theory of a defect in the street, in action for damages due to injuries to travelers from awnings, signs, billboards, poles, electric wires, or other objects suspended over, or near thereto, or falling into a street or sidewalk.[86]

Courts considered any invasion of the airspace above a public right of way to constitute a nuisance.[87] Even a bay window on a second story, 16 feet above the ground, which extended approximately three feet into the line of the street was held a public nuisance.[88] So, too, a wire temporarily erected as part of a Fourth of July celebration was considered an obstruction of the street below.[89] Awnings, bridges, trees, bay windows—anything projecting into the space above public land, without the explicit approval of the municipality, constituted a de facto nuisance.[90]

Public right of way similarly extended downward, into the earth.[91] While certainly a duty to keep the streets safe was operable (thus implicating nuisance), state courts routinely looked to the city or town *qua* property owner also to consider such invasions of airspace a trespass. The Supreme Court of Iowa, for example, noted in relation to a wire mounted by an electric company above the road:

> The city being the owner in fee simple of the streets, of necessity its rights extend above the surface thereof. How far, we need not determine in this case; but, since it is entitled to the absolute control and occupancy of the space above these streets, any invasion thereof by stretching wires thereon at this height of necessity is an infraction of the rights of the city, and amounts to a trespass.[92]

STATE CONTROL OF PROPERTY RIGHTS

The Articles of Confederation acknowledged that each state retained "its sovereignty, freedom, and independence, and every Power, Jurisdiction and right, which is not by this confederation expressly delegated to the United States, in Congress assembled."[93] Over the next decade, the lack of a sufficiently robust national government forced a renegotiation of the terms of the union. Yet even as delegates drafted the Constitution, they preserved state sovereignty to ensure a check on federal power.

For some, the document did not go far enough—a concern that the Federalist Papers addressed. In them, James Madison argued that states retained their powers. Each branch of the national government would "owe its existence more or less to the favor of the State governments, and must consequently feel a dependence, which is much more likely to beget a disposition too obsequious than too overbearing towards them."[94] Alexander Hamilton scoffed at the idea that states would become subservient to federal regulation, calling such speculation "idle and visionary."[95] The Constitution would guard against federal overreach by limiting the national government to the enumerated authorities, in sharp contrast to the status of states as governments of general jurisdiction. Madison explained:

> The powers delegated by the proposed Constitution to the federal government are few and defined. Those which are to remain in the State governments are numerous and indefinite. The former will be exercised principally on external objects, as war, peace, negotiation, and foreign commerce. . . . The powers reserved to the several States will extend to all the objects which, in the ordinary course of affairs, concern the lives, liberties, and properties of the people, and the internal order, improvement, and prosperity of the State.[96]

Congress could not act outside of the powers specified in Article I, § 8. Even within the enumerated powers, the capital could not be established without the states affirmatively agreeing to cede territory.[97] The Constitution forbade federal acquisition of state territory, or the partitioning of existing states, without their legislature's consent.[98] The Constitution established—and explicitly recognized—dual sovereignty. States, consistent with the Guarantee Clause, constituted republics in their own right.[99] Their citizens were to enjoy the privileges and immunities of their sister states.[100]

States retained all powers neither delegated to the federal government nor prohibited to them.[101] The powers to which the Tenth Amendment referred were extensive. Police powers (relating to health, welfare, and morals), criminal law, and corporate charters fell within the state domain. States had authority over education, manufacturing, and agriculture. They were responsible for regulating, controlling, and governing real and personal property, as well as individuals within state borders. As the Supreme Court explained in 1905:

> Although this court has refrained from any attempt to define the limits of [state police powers], yet it has distinctly recognized the authority of a state to enact quarantine laws and health laws of every description; indeed, all laws that relate to matters completely within its territory and which do not by their necessary operation affect the people of other states.[102]

A dozen years later, the Court again acknowledged that while it had not tried to define police powers, "its disposition is to favor the validity of laws relating to matters completely within the territory of the state enacting them."[103]

What this meant was that, as a practical matter, control of real property and associated questions fell to state and local governments. State sovereignty was so strong that federal power only extended to property exclusively in federal control. Thus, one of the first efforts by Congress to outlaw murder or robbery in any river, basin, or bay fell outside constitutional bounds. As the U.S. Supreme Court explained in 1818, the effort by the government to prosecute a murder in Boston harbor on board the ship *Independence* under federal law ran afoul of the powers retained by the states.[104]

State cases followed course. In *Commonwealth v. Young*, also decided in 1818, the Pennsylvania Supreme Court explained, "The legislative power and exclusive jurisdiction remained in the several states, of all territory within their limits, not ceded to, or purchased by, congress, with the assent of the state legislature, to prevent the collision of legislation and authority between the United States and the several states."[105] Any real property not owned by the federal government was under state jurisdiction.

The position of a state was the same as that of a sovereign nation, which owed a duty to its citizens to act in the manner that best advanced their interests. In 1837, the U.S. Supreme Court acknowledged,

> [A] State has the same undeniable and unlimited jurisdiction over all persons and things, within its territorial limits, as any foreign nation; where that jurisdiction is not surrendered or restrained by the Constitution of the United States. . . . [B]y virtue of this, it is not only the right, but the bounden and solemn duty of a State, to advance the safety, happiness and prosperity of its people, and to provide for its general welfare . . . [A]ll those powers which relate to merely municipal legislation, or what may, perhaps, more properly be called *internal police*, are not thus surrendered or restrained; and that, consequently, in relation to these, the authority of a State is complete, unqualified and exclusive.[106]

When Alabama joined the union in 1819, the federal government premised its admission on the condition that its navigable waters would remain public highways.[107] The Supreme Court circumscribed the requirement, noting that it did *not* deprive the state of its rights over navigable waters; nor did it affect state control over the shores or the soil under the navigable waterways. "[T]he United States never held any municipal sovereignty, jurisdiction, or right of soil in and to the territory, of which Alabama or any of the new States were formed," the Court observed.[108] The state therefore was "entitled to the sovereignty and jurisdiction over all the territory within her limits, subject to the common law."[109] The national government could not demand that the state cede its authority. Such a power would run rampant over the very concept of federalism.

The national government's authority undergirding the Alabama provision was rooted in Congress's Commerce Clause powers, which were initially narrowly interpreted.[110] In *Gibbons v. Ogden*, the Court made it clear that while the navigation of waterways impacted interstate commerce, interior state traffic lay outside its remit:

> The genius and character of the whole government seem to be, that its action is to be applied to all the external concerns of the nation, and to those internal concerns which affect the States generally; but

> not to those which are completely within a particular State, which do not affect other States, and with which it is not necessary to interfere, for the purpose of executing some of the general powers of the government. The completely internal commerce of a State, then may be considered as reserved for the State itself.[111]

Expressio unius exclusio alterius, by demarcating among the states, what lay within them was beyond Congress's purview.[112] Inspection laws, too, lay outside federal control, as they fell within "that immense mass of legislation which embraces everything within the territory of a State, not surrendered to the general government: all which can be most advantageously exercised by the States themselves."[113] Chief Justice John Marshall explained:

> Inspection laws, quarantine laws, health laws of every description, as well as laws for regulating the internal commerce of a State, and those which respect turnpike roads, ferries, &c., are component parts of this mass. No direct general power over these objects is granted to Congress; and, consequently, they remain subject to State legislation.[114]

While the federal government might control navigable waterways insofar as interstate commerce went, it operated within the strict limits long recognized as part of common law. For centuries riparian water rights lay with those landowners whose property abutted navigable waterways. Common law recognized their right to access water for transportation, building, and hunting or fishing. Congress has acknowledged states' title to submerged navigable lands.[115]

At the Founding, the authority for regulating property rights transferred to the states. Landowners' title to the water extended to the low-water mark, while the ground beneath was held *by the state*, not the federal government. This principle is well recognized by the courts.[116] States own the beds under navigable waters, while adjacent landowners, absent other legal arrangements, own the water above. States—not the federal government—control and place any applicable restrictions on the landowners in their use and enjoyment of the property. Thus, an ordinance limiting the height of buildings in Baltimore and requiring that property owners obtain a permit before building, altering, or repairing any structure within city limits, fell within state police powers.[117] In New Hampshire, a statute forbidding the erection of fences higher than five feet when the purpose was to annoy adjoining owners or occupants similarly fell within the general powers of the state.[118]

Real estate law is governed by the general principles of contract law and individual state law. Deeds are executed and delivered to state entities. Building regulations, placement of billboards and advertising, zoning, and other matters impacting real property are handled by states and, in many cases, local municipalities. The Supreme Court has regularly upheld state and local authority in these areas.[119] Courts only subject municipal zoning ordinances to rational basis review.[120] They give deference to the states for the rules regulating things that extend upward into the air, such as billboards.[121] In considering the constitutionality of one such ordinance, the Court noted that while it had thus far refrained, "from any attempt to define with precision the limits of the police power . . . its disposition is to favor the validity of laws relating to matters completely within the territory of the state enacting them."[122]

Outside of property actually held by the federal government, it is the states who control property rights from the ground up—not the federal government. And states, as well as landowners, are limited by *ad coelum* in what they can do in regard to real property.

AVIATION LAW

By the time Pollock's 11th edition of the *Law of Torts* was published in 1920, aircraft were becoming ever more widespread. According to the *New York Times,* 2,200 airplanes were in commercial use at that time and around 500 more were owned privately, with 92 companies operating planes and nearly twice as many manufacturing aircraft.[123] Numerous calls were being made, on behalf of industry, to adopt federal laws to support the nascent industry.[124] The American Bar Association and the National Conference of Commissioners on Uniform State Laws put committees together that year to determine what laws should govern airspace.[125]

Early commentaries on how the law should evolve all questioned whether the federal government had the constitutional authority to pass any measures regulating aviation. The ancient recognition of *ad coelum* appeared to foreclose federal jurisdiction.[126] As a judicial matter, consistent with *ad coelum*, cases dealing with air travel had come down clearly on the side of the property owner. Thus, in 1822, when an aeronaut landed a balloon and damaged property, the aeronaut was held liable.[127] In 1912, an aviator flying 100 feet off the ground who did not cause any damage to property was likewise liable for trespass.[128] Common law applied. The theory underlying these and other cases was that title carried with it the right to occupy, use, and possess the airspace above the ground.

To address the constitutional concerns that dogged federal initiatives, it was initially proposed that the admiralty clause conferred the necessary authority.[129] Courts, however, rejected this reasoning, as air flight was more akin to the land than the sea.[130] Others invoked the war power clause, which was roundly rejected on the grounds that adopting this approach would create limitless federal power.[131] Nor could the constitutional authority to make treaties provide the grounds for federal measures. The United States had not actually ratified the International Air Navigation Convention of 1919 (see discussion later), and it was questionable whether an agreement dealing exclusively with international travel had any bearing on interstate matters.[132] In the interim, the powers conveyed by the commerce clause, as discussed previously, were narrowly defined.

The constitutional concerns prompted the American Bar Association committee that had been formed to look into the matter to call for a formal amendment to bring air travel within federal control.[133] This perspective was shared by the Federal Air Service, whose legal adviser proposed either a constitutional amendment or the purchase of air avenues (on account of such travel amounting to a taking) to make it legal.[134]

UNIFORM AERONAUTICS ACT

Absent a constitutional amendment, as air travel proliferated and safety concerns mounted, the National Conference of Commissioners on Uniform State Laws prepared and, with the American Bar Association's approval, recommended that states adopt the Uniform Aeronautics Act.[135] The model statute focused on state sovereignty over airspace, landowners' rights over the air above their property, the lawfulness of flight, damage and collision, jurisdiction, and the use of aircraft. In so doing, the model statute clearly established state dominion.[136] The act provided that the ownership of the air above the land and waters of a state is vested in the several owners of the surface beneath, subject to certain overflight rights.[137] It recognized state contract, tort, and criminal jurisdiction over pilots, passengers, and aircraft when in flight over the state.[138] The model statute imposed absolute liability on the owner of every aircraft for injuries to persons or property on land or the water beneath, caused by an aircraft or any object falling from it.[139] Except for contributory negligence of the property owner or the person harmed, liability lay with the owner of the aircraft regardless of whether the owner was negligent.[140]

Twenty-two states adopted the Uniform Aeronautics Act in some form.[141] In Massachusetts, for instance, General Law 534 of 1922 established a licensing and registration system for flight.[142] The law established control over private and commercial aircraft takeoff, landing, and flight.[143] State law also provided that aircraft should not be operated "over any thickly settled or business district at an altitude of less than 3,000 feet, or over any building or person at an altitude of less than 500 feet, except when necessary for the purpose of embarking or landing." This law was tested in the 1930 case of *Smith v. New England Aircraft Co.*[144] The Supreme Court of Massachusetts, in an opinion issued by Chief Justice Arthur Rugg, considered the question of whether overflight constituted a trespass to be well within the state domain:

> It is essential to the safety of sovereign States that they possess jurisdiction to control the airspace above their territories. It seems to us to rest on the obvious practical necessity of self-protection. Every government completely sovereign in character must possess power to prevent from entering its confines those whom it determines to be undesirable. That power extends to the exclusion from the air of all hostile persons or demonstrations, and to the regulation of passage through the air of all persons in the interests of the public welfare and the safety of those on the face of the earth. This jurisdiction was vested in this Commonwealth when it became a sovereign State on its separation from Great Britain.[145]

At the Founding, common law conveyed control over the air to the states as an aspect of their sovereignty. Subsequent statutes regulating the operation of aircraft were "enacted under the police power."[146] That authority can be used to regulate private rights. The court explained, "There are numerous statutes upheld as an exercise of the police power interfering with, narrowing and regulating, private rights of landowners in the use of their estates."[147] Fishery rights, building regulations, the setback for homes, the installation of sprinklers, prohibitions on building heights, and zoning districts, *inter alia*, fall into this category.

State law, "by plain implication, if not by express terms," recognized and authorized "the flying of aircraft over privately owned land." Air travel had become an important part of life. Nevertheless, even "after making every reasonable legal concession to air navigation as commonly understood and as established under the statutes," the facts of the case, "under settled principles of law" (i.e., *ad coelum*) constituted "trespass to the land of the plaintiffs so far as concerns the take-offs and landings at low altitudes and flights."[148] For the court, "ownership of airspace extends to reasonable heights above land." Thus, "Air navigation, important as it is, cannot rightly levy toll upon the legal rights of others for its successful prosecution."[149] At higher levels, unoccupied airspace could be used, but under a certain level, the ancient rights of property owners prevailed, making any interference in the airspace a trespass.

In *Smith v. New England Aircraft Co.*, a 92-acre airport abutted the plaintiff's land, which was a mostly wooded country estate of some 270 acres. Flights taking off from the airport traversed the property less than 100 feet off the ground, but not less than 500 feet over any buildings. Pilots leaving the airport followed the statutory provisions laid out by the state. For Chief Justice Rugg, the noise did not appear to materially interfere with the plaintiff's physical comfort; nor did the flights subject any member of the household to fear. There had been no damage of any sort. "I find," the chief justice wrote, "the plaintiffs are persons accustomed to a rather luxurious habit of living, and while the noise from the airplanes in flight over their premises has caused them irritation and annoyance . . . gauged by the standards of ordinary people this noise is not of sufficient frequency, duration or intensity to constitute a nuisance."[150]

THE AIR COMMERCE ACT OF 1926

In 1922, the first bill proposed in Congress to govern civil air navigation failed to pass.[151] Four years later, the 1926 Air Commerce Act became law.[152] Its purpose was to encourage and to support interstate and international commerce by supporting the nascent air industry.[153] Accordingly, it directed the secretary of commerce to ensure that new restrictions or regulations did not unduly hamper innovation.[154] The law charged the secretary with figuring out how to capitalize on aeronautics, to issue and enforce rules to make flight possible, to provide a licensing regime for airmen, to certify aircraft, and to establish airways.[155] The statute made bureaucratic changes to support its edict, establishing a new aeronautics branch at the Commerce Department, headed by an assistant secretary.[156]

The Air Commerce Act of 1926 acknowledged dual sovereignty and the rights of state governments. It was careful to distinguish between interstate and international travel, which fell within Congress's Commerce Clause authorities, and intrastate activities, which fell outside congressional reach. It thus authorized the secretary of commerce to establish the requisite navigation facilities and airways, but not airports, which firmly fall within state and local control.[157] Only aircraft in navigable airspace, and therefore in interstate lanes, would be subject to regulation. The statute defined navigable airspace as "airspace above the minimum safe altitudes of flight prescribed by the Secretary of Commerce," which would be "subject to a public right of freedom of interstate and foreign air navigation."[158] This distinction preserved for the states (and via them, landowners) control over the ground and the air above it.

The House report accompanying the bill drew parallels with marine navigation.[159] It noted that Congress had drawn directly from the registration and inspection laws introduced in 1789 and 1838 for vessel registration and inspection.[160] The aids to navigation supplied by the federal government, moreover, would align with the documents issued by the Department of Commerce's Bureau of Lighthouses and U.S. Coast and Geodetic Survey, providing for fluidity in interstate and international navigation.[161] In sum, "The whole framework and, in many cases, the very language of the bill may fairly be said to be merely the application to air transportation of provisions of statutes and principles of law long established as to water transportation."[162] As with navigable waterways, the Commerce Clause provided the constitutional hook.[163] Only public flight in the navigable airspace would be subject to federal control.[164]

The Commerce Clause claim reflected the position of the United States in the international system. As an aspect of sovereignty, each country had the right to complete and exclusive control over the airspace above its territory—a right recognized in the 1919 International Convention Relating to the Regulation of Aerial Navigation.[165] The United States never ratified the convention, which was later superseded.[166] Nevertheless, Congress cited it in support of the 1926 act. Under the convention, sovereign control included not just the national territory of the mother country and colonies, but also the adjacent territorial waters.[167]

Just two years after Congress passed the Air Commerce Act, the Inter-American Commercial Aviation Convention reiterated the importance of protecting the sovereignty of each country over its airspace.[168] Sovereign control explicitly included the "right to prohibit, for reasons which it deems convenient in the public interest, the flight over fixed zones of its territory by the aircraft of the other contracting states and privately owned . . . aircraft."[169] International air travel was organized along national lines: like ships, aircraft carried the nationality of the country in which they were registered, which had to be conveyed through special markings.[170]

The right to self-defense underlay sovereign control. Accordingly, in 1929 the Protocol to the International Convention Relating to the Regulation of Aerial Navigation established:

> Each contracting State is entitled for military reasons or in the interest of public safety to prohibit the aircraft of the other contracting States, under the penalties provided by its legislation and subject to no distinction being made in this respect between its private aircraft and those of the other contracting States, from flying over certain areas of its territory.[171]

Under the protocol, states could prohibit other countries' aircraft from flying over their territory during peacetime as well.[172] Presaging technologies that have only come to fruition with the advent of drones in the 21st century, the protocol had a special provision applicable to unmanned aircraft: "No aircraft of a contracting State capable of being flown without a pilot shall, except by special authorization, fly without a pilot over the territory of another contracting State."[173]

Following passage of the 1926 Air Commerce Act, the secretary of commerce promulgated new regulations that established an absolute floor of 1,000 feet over cities, towns, and settlements, and 500 feet over all other land "except where indispensable to an industrial flying operation" as a basis for determining navigable airspace.[174] These regulations acknowledged state and landowner control beneath that height.

A number of states adopted parallel provisions. In 1927, for instance, Wyoming defined navigable airspace as "the airspace above the minimum altitudes of flight which are hereby defined to be not less than 1,000 feet over any city, town, or settlement, and not less than 500 feet over any other portion of the state of Wyoming except in case of landing, taking off, or emergencies necessitating lower flight, and excepting lower flight when necessary for industrial operations."[175] In Ohio, the General Act Relative to Aeronautics similarly brought state law into line with the contours of the federal regulations, even as it recognized that it was within the state's purview whether or not to do so.[176] The statute made clear that for policy reasons (uniform regulations and public safety), the state was *choosing* to align its measures with those of the federal government.

In 1930, Ohio's provisions were challenged.[177] The case was brought by R. H. Swetland, who in 1905 had bought 135 acres in a rural area to be used as a country estate. The area was still sparsely settled farmland when in 1929, a Curtiss-Wright subsidiary bought 272 acres directly across the road from Swetland's property to operate an airport and flying school. Swetland immediately responded: the deed ceding the property to Ohio Air Terminals was dated May 23, 1929, and recorded five days later. Swetland sent a letter protesting the purchase on May 29, with legal action commencing June 1. Undeterred, Curtiss Airports built a 20-plane hangar and planned a parking lot for 460 cars.

The question before the court in *Swetland v. Curtiss Airports Corporation* was whether the Ohio state law in question was a reasonable exercise of the police power, as well as whether *ad coelum* established such property rights in a landowner as to make overflight either a nuisance or a trespass.[178] Looking to the state laws facilitating the growth of air travel, the court took judicial notice of the fact that "although aviation is still to some extent in the experimental stage, it is of great utility in times of peace, and will be a great protection to the nation in times of war.

In fact, it is indispensable to the safety of the nation that airports and flying schools such as contemplated by the defendants be encouraged in every reasonable respect."[179] With state measures regulating it, the only way the airport could be a nuisance was if it was "located in an unsuitable location" or was "operated so as to interfere unreasonably with the comfort of adjoining property owners."[180] While the warm-up of airplanes and their flight over the neighbors' property might be noisy, it was not to "such a degree as to annoy persons of ordinary sensibilities."[181] Nor was dust an issue, because the airport could simply install concrete runways or use runways with a sufficient amount of grass.[182] Crowds did not present a concern. The court had refused to enjoin amusement parks, for instance, merely on those grounds.[183] Once the novelty had worn off, moreover, fewer people would come. Dropping leaflets or putting up lights might be a nuisance, but such actions could be enjoined as necessary.[184]

Having rejected the airport as constituting a nuisance, the court turned to allegations of trespass. "An owner's rights in land in this state are amply protected by constitutional guaranties," the court wrote, "but what those rights are, so far as air space above the land is concerned, has not been declared by legislation, nor have such rights been fixed by the courts. That the landowner's rights are not limited to the surface of the earth, but extend into the space above it, is settled by many well-considered cases."[185]

The court addressed *ad coelum*: "The venerability of this maxim, its frequent repetition, and the high standing of many of those who have relied upon it, not only warrant, but call for, a careful consideration of its origin and application in adjudicated cases."[186] In none of the cases, though, had the maxim been applied explicitly to air travel. The salient question was at what point the *coelum* was reached. According to Professor Hiram Jome, the clause, as employed by Latin writers meant the lower airspace, where birds fly and clouds drift, rain falls and lightning strikes.[187] "Occasionally," Jome wrote, "it meant God, 'heaven the home of the happy dead,' and the resting place of the stars."[188] Consulting treatise writers on torts law, the court noted that a number of them recognized ownership in airspace appurtenant to land, up to a reasonable height.[189] The court could not find any constitutional or legislative provisions or statutes that had established the exclusive proprietary right "in a landowner to the superincumbent air space normally traversed by the aviator."[190]

Because the establishment of a 500-foot minimum altitude rule seemed to be a reasonable exercise of federal power—and one with which the Ohio legislature appeared to agree, flight at that height was legal. Below that, however, it was "conceivable and very probable" that pilots could not interfere with the landowner's "effective possession" of the airspace.[191]

On appeal, the U.S. Court of Appeals for the Sixth Circuit observed that "the law reports of practically every state" refer to *ad coelum*; however, the right was not absolute: "we cannot hold that in every case," the court wrote, "it is a trespass against the owner of the soil to fly an aeroplane through the air space overlying the surface."[192] Nevertheless, "[t]his does not mean that the owner of the surface has no right at all in the air space above his land. He has a dominant right of occupancy for purposes incident to his use and enjoyment of the surface, and there may be such a continuous and permanent use of the lower stratum which he may reasonably expect to use or occupy himself as to impose a servitude upon his use and enjoyment of the surface."[193] The landowner could not "reasonably expect to occupy" the upper stratum. Here, the only right of the landowner was "to prevent the use of it by others to the extent of an unreasonable interference with his complete enjoyment of the surface."[194] While interference amounted to a trespass in the lower strata, in the upper strata, only an action for nuisance could lay. The court was unable to "fix a definite and unvarying height below which the surface owner may reasonably expect to occupy the air space for himself."[195] Such a designation would be fact dependent. For this case, it was sufficient to say that the defendants were in the upper strata, above the 500-foot floor. No trespass, therefore, had occurred. As a matter of nuisance, however, the airport had already interfered with the family's enjoyment of the property and depreciated its value.

In his concurrence, Judge Smith Hickenlooper, a Calvin Coolidge appointee, rejected the idea that trespass could only occur at the lower strata. Higher up, "although a single flight over the plaintiff's land may not constitute a trespass, such flights may be so continuous as in the aggregate to do so."[196] As a logical matter, for the aggregate of a large number of flights to constitute a trespass, "it must be because each of the said flights is itself a trespass."[197]

With property rights firmly in the hands of the states, a Uniform Licensing Act was proposed in 1930.[198] The National Conference of Commissioners on Uniform State Laws supplemented the statute in 1935 with a Uniform Aeronautical Regulatory Act. The model law made it mandatory for all pilots to hold federal licenses and added state supervisory authority to license airports, educational institutions, clubs, and other associated areas.[199] It further provided for municipalities and counties to acquire, construct, and regulate airports.[200] While not adopted *in toto* by all states, many of the provisions of these acts found their way into state law.[201]

States had the lead. *Cujus est solum, ejus est usque ad coelum et ad inferos* was considered by scholars "one of the most firmly and permanently established rules of our common law."[202] Nevertheless, the federal government continued to push for control of airspace. As a consequence, the early 1930s saw a sudden spate of scholarship on the implications of air travel for the historical protection of property owners over the airspace adjacent to their land, as well as for the federalism divide.[203] It also prompted the creation of a new journal, the *Journal of Air Law and Commerce*, whose editorial stance rather weighted the deck in favor of the feds.

CIVIL AERONAUTICS ACT OF 1938

The Civil Aeronautics Act of 1938 marked a significant shift in federal control. The statute was largely driven by safety concerns. At that point, some 17,681 licensed civil pilots were flying 7,300 licensed civil airplanes between 2,327 airports.[204] The commercial aspects of flight were beginning to take off: 1,250,000 passengers per year and over 1.25 billion pound-miles of mail per month were being transported by air.[205] Simultaneously, in what appeared to be a race to the bottom, as competition increased, carriers tried to cut costs, potentially undermining passenger safety.[206] Following a series of harrowing air incidents, the Senate directed the Committee on Commerce to investigate and to report on how best to improve air safety.[207] In debating the resolution, one senator explained, "[T]his bill will not only promote an orderly development of our Nation's civil aeronautics, but by its immediate enactment prevent the spread of bad practices and of destructive and wasteful tactics resulting from the intense competition now existing within the air-carrier industry."[208] Regulating the air industry appeared, from a federal perspective, to be no different from regulating other public utilities.

Congress proclaimed in the 1938 statute,

> The United States of America is hereby declared to possess and exercise complete and exclusive national sovereignty in the air space above the United States, including the air space above all inland waters and the air space above those portions of the adjacent marginal high seas, bays, and lakes, over which by international law or treaty or convention the United States exercises national jurisdiction.[209]

This statement went considerably beyond the 1926 act, which had claimed sovereignty vis-à-vis *other countries*, but not as against the states.[210] The Supreme Court later rejected Congress's claim, noting that the 1926 statute had not "expressly exclude[d] the sovereign powers of the states."[211] To the extent that Congress was laying claim to matters within state jurisdiction, it was intruding on a constitutionally protected area. States retained control over the airspace immediately above state land and waters. No less than three provisions of the uniform act adopted by the states had asserted this authority.[212]

The Civil Aeronautics Act laid out a complex structure to ensure air safety. It established regulations for licensing pilots, aircraft, and air carriers and for maintaining equipment.[213] It created an Air Safety Board and incorporated civil and criminal penalties for failing to meet the requirements laid out in the act.[214] It introduced a range of regulations applicable to both domestic and foreign air carriers and the transportation of mail.[215] The act transferred responsibility from the Bureau of Air Commerce to a new, independent agency, the Civil Aeronautics Authority (CAA), which was tasked with regulating air transportation, encouraging foreign and domestic commerce, improving the mail service, and providing for matters related to national defense.[216] From 1938 to 1939, the CAA examined 2,668 accidents and made recommendations.[217] The agency went on to analyze passenger flow and the need for new routes, airmail pricing, and transatlantic trade, and to participate in international discussions about aviation regulation.[218] The CAA also operated a Bureau of Safety Regulations, which took on pilot and mechanic certification, aircraft inspection, and other matters.[219] It further ran a training program, coordinating with 435 colleges and 528 flight schools to train and certify some 8,313 new pilots.[220] Over the course of fiscal year 1940, there were zero aviation-related fatalities across the United States.[221]

States continued to serve as the lead for control of the land. Accordingly, in 1939 the CAA, together with the National Institute of Municipal Law Officers, wrote a Model Airport Zoning Act for the states.[222] It underwent several revisions, with the aim, as one scholar explains, of empowering states

> to promulgate, administer and enforce under the state police power, airport zoning regulations limiting the height of structures and objects of natural growth and otherwise regulating the use of property in the vicinity of public airports, and to acquire by purchase, grant or condemnation, air rights and other interests in land, for the purpose of preventing obstruction of the airports' approaches.[223]

By 1947, most states had adopted the measures. The partnership between states and the federal government took further form with the 1946 Federal Airport Act, which provided a $500 million grant over a seven-year period to states and municipalities interested in building airports.[224] In 1950, the act was extended to 1958, with funding provided for runways and taxiways.[225] Further laws extended funding until the act was finally repealed and replaced by the Airport and Airway Development Act of 1970.[226] In 1958 the CAA became the Federal Aviation Agency (later the Federal Aviation Administration).[227]

OF NUISANCE AND TRESPASS

Aircraft traveling over farmland presented a particular risk of stress on livestock. This issue came to the fore in Nebraska fairly early on in *Glatt v. Page*.[228] The court prohibited any flights under 500 feet over the plaintiff's house and under 250 feet over adjoining fields outside of takeoff and landing. The court was careful to note the special risks at that time, as "flying . . . at an altitude of less than 100 feet will ordinarily frighten teams at work . . . and thereby render cultivation of plaintiff's land in that vicinity difficult and hazardous." Frequent flight at this level "constitutes a damage to the plaintiffs and an impairment of the use and enjoyment of said premises which goes with the land and belongs to plaintiffs as well as the soil thereof."[229]

In 1946, the Supreme Court came to a similar conclusion, recognizing that, as a corollary to landowner rights *ad coelum*, repeated use of the airspace above a farmer's property for takeoff and landing constituted a Fifth Amendment taking. In *United States v. Causby*, despite the federal government's bald assertion of "exclusive national sovereignty" and the right of freedom in air transit, the Court came down on the side of the landowner's property rights.[230] A chicken farm, located near an airport routinely used by the military, essentially had been subjected to a servitude. While *cujus est solum ejus est usque ad coelum* recognized ownership to the periphery of the universe, it proved ill-fitting in the modern world.[231] Air above the minimum safe altitude of flight, like a public highway, amounted to the public domain. However, anything below the navigable airspace was not. It was in the full control of the owner: "The landowner owns at least as much of the space above the ground as he can occupy or use in connection with the land. The fact that he does not occupy it in a physical sense—by the erection of buildings and the like—is not material."[232] The Court explained, "The superadjacent airspace at [a] low altitude is so close to the land that continuous invasions of it affect the use of the surface of the land itself. We think that the landowner, as an incident to his ownership, has a claim to it and that invasions of it are in the same category as invasions of the surface."[233] The immediate reach of adjacent airspace thus belonged to the landowner.

The Court recognized that the problem of granting the federal government control over the lower airspace would give them "complete dominion and control over the surface of the land."[234] This would eviscerate federalism as well as property rights.

Following *United States v. Causby*, Congress redefined "navigable airspace" to mean "airspace above the minimum altitudes of flight prescribed by regulations issued under this chapter," including "airspace needed to insure [sic] safety in take-off and landing of aircraft."[235]

In 1962, the Supreme Court had to ascertain whether, under new definition, takeoff and landing areas abrogated property rights.[236] In *Griggs v. Alleghany*, the Supreme Court again recognized that landowners own the airspace above their property: "[T]he use of land presupposes the use of some of the airspace above it. Otherwise, no home could be built, no tree planted, no fence constructed, no chimney erected. An invasion of the superadjacent airspace will often affect the use of the surface of the land itself."[237] The imposition of a glide path above someone's property amounted to a taking. Any exercise of dominion over adjacent airspace would have to be compensated.

CONCLUDING REMARKS

With the history of *ad coelum* and state police powers in mind, one could be forgiven for being surprised by contemporary federal claims to airspace related to the proliferation of UAS technologies. Navigable airspace, as an anchor for expanded federal control, cannot extend to the ground without violating property rights and state sovereignty.

This does not mean that Congress has no role to play. To the extent that unmanned aircraft of any sort travel 500 feet above the ground, they would be subject to the provisions governing other aircraft. So, too, would federal control extend to the safety of aircraft ascending and descending from airports. In 2012, for instance, the FAA Modernization and Reform Act prohibited the FAA from regulating model aircraft weighing less than 55 pounds, unless they were flown within five miles of an airport.[238]

But the federal government is now attempting to bring the floor of navigable airspace ever lower: under the FAA Reauthorization Act of 2018, the hobbyist carveout is limited to 400 feet above ground level.[239] The FAA is placing restrictions on flying drones at night and requiring users to register their drones and fly within visual line of sight. In addition, the FAA now imposes flight restrictions on recreational drone operators—such as when Pope Francis visited Philadelphia in 2015.[240]

To the extent that the FAA is attempting to cast its net more widely, property rights and state authority present an important limitation. Numerous states already have asserted their constitutional power to regulate drones.[241] Some states, such as California and Louisiana, do not put any ceiling on the air above private property.[242] Any entry is considered a trespass. Where ceilings *have* been set, they range from 250 feet above the ground (Nevada) up to 500 feet above property (Tennessee)—a number reflecting the navigable airspace, as established in the Code of Federal Regulations. Tennessee code defines criminal trespass to include an unmanned aircraft entering the portion of airspace above an owner's land not regulated by the FAA as navigable airspace.[243] In Nevada, the state legislature has prohibited flying a drone within 500 feet of any critical infrastructure.[244] A small unmanned aircraft flying lower than 500 feet does not expand the federal government's jurisdiction.

The FAA is now contemplating regulation of commercial drones that may fly over public roads—property held by the state. In 1959 the Supreme Court recognized the importance of state sovereignty, even in the face of Commerce Clause claims:

> The power of the State to regulate the use of its highways is broad and pervasive. We have recognized the peculiarly local nature of this subject of safety, and have upheld state statutes applicable alike to interstate and intrastate commerce, despite the fact that they may have an impact on interstate commerce.[245]

The Court had long recognized state power over public highways. In 1938, the Supreme Court considered a South Carolina statute regulating the size of vehicles that could use its roads.[246] The trial court had determined that the statute unreasonably burdened interstate commerce.[247] The Supreme Court saw it rather differently: "South Carolina has built its highways and owns and maintains them. While the constitutional grant to Congress of power to regulate interstate commerce has been held to operate of its own force to curtail state power in some measure" it does not so impact *all* actions affecting interstate commerce.[248] The Court noted that since 1829:

> [I]t has been recognized that there are matters of local concern, the regulation of which unavoidably involves some regulation of interstate

> commerce but which, because of their local character and their number and diversity, may never be fully dealt with by Congress. Notwithstanding the commerce clause, such regulation in the absence of congressional action has for the most part been left to the states by the decisions of this Court.[249]

In the context of regulating vehicles traveling along the ground, there was nothing to indicate that the states would have to "curtail[] their power to take measures to insure the safety and conservation of their highways which may be applied to like traffic moving intrastate."[250]

Presumably, the Court will take the same view of drones. Like small unmanned aerial systems, vehicles, and the routes they traveled, mattered: "[f]ew subjects of state regulation are so peculiarly of local concern as is the use of state highways."[251] States play the primary role in building them. They own them. They maintain them. "The state," moreover, "has a primary and immediate concern in their safe and economical administration."[252] States have an even greater interest in things that travel through the air and impact the people and vehicles below. Virtually any regulation of the roads (or the air) could affect interstate commerce—but this did not, in any way, diminish the strong state interest in regulating its own highways. For the Court, South Carolina was well within its constitutional power to act.[253]

The Court came to the same conclusion about a Pennsylvania statute that prohibited any cars carrying other vehicles over the head of the vehicle operator.[254] The Court explained that highways "are state owned, and, in general, are open in each state to use by privately owned and controlled motor vehicles of widely different character."[255] More sensitive than the federal government to local needs and conditions, states had long been in charge of the safety of their roads.[256] The effort by the federal government to assert national control raised "problems of peculiar difficulty and delicacy."[257] Ensuring road safety constituted a legitimate exercise of state police power.[258]

Just because the federal government has acted does not divest states of their core authority. As the Supreme Court explained in yet another case dealing with a state motor vehicle provision:

> An examination of the acts of Congress discloses no provision, express or implied, by which there is withheld from the state its ordinary police power to conserve the highways in the interest of the public and to prescribe such reasonable regulations for their use as may be wise to prevent injury and damage to them.[259]

For the Supreme Court, the regulation of state highways was "akin to quarantine measures, game laws, and like local regulations of rivers, harbors, piers, and docks, with respect to which the state has exceptional scope for the exercise of its regulatory power, and which, Congress not acting, have been sustained even though they materially interfere with interstate commerce."[260]

For cases coming to the Court, state control over use of land within its bounds is presumed to be constitutionally valid—even if it interferes with interstate commerce, unless there is an *excessive* burden placed on it.[261] In 2002, a parallel effort to establish federal preeminence over states in regard to aerial advertising failed. Skysign International, which was operating under an FAA certificate of waiver, flew over densely populated areas in violation of a municipal ordinance.[262] Prior to introducing the ordinance, Honolulu had twice asked the FAA whether the federal measure preempted its ability to regulate the airspace and had been informed, on both occasions, that it did. The FAA, however, had been mistaken. The Ninth Circuit, ruling for Honolulu, observed that advertising is one of those areas that traditionally falls within the state domain. The federal government could not displace that authority without a clear statement that it intended to do so—which could itself be challenged.

Congress's claim to sovereignty over U.S. airspace as an instrument of international control does not eviscerate domestic divisions. Instead, as the Court in *Braniff* noted, it is merely "an assertion of exclusive national sovereignty."[263] States and property owners (via state law) control the airspace above the land. Should federal power be understood as extending to everything above land and roads, it would so undermine state police powers over property as to render them virtually nonexistent. Such an approach flies in the face of ad column and state sovereignty, long considered at the core of U.S. law.

CHAPTER SIX

LEGISLATIVE RULES FOR USE OF DRONES BY LAW ENFORCEMENT

JAKE LAPERRUQUE

In recent decades governments at the federal, state, and local levels have deployed a wide array of invasive surveillance technologies. Often these tools and techniques aren't restricted by any clear social norms, and frequently they upend long-standing ideas, such as "you can't have a right to privacy in public." Virtually all of these new types of surveillance also came into use absent legal rules and restrictions. In response, legislatures and courts have scrambled—to the extent that such slow-moving bodies can—to put proper rules in place and stop the 21st century from becoming a Wild West of surveillance. Policymakers' response to drone surveillance, a premier example of a disruptive and rapidly advancing surveillance tool, should be centered on four overarching rules: a probable cause requirement, an exhaustion requirement, minimization rules, and a rule requiring governments to provide logistical information with a justification for the extent of surveillance they request.

Drones were originally deployed mostly by the military. Reconnaissance drones—capable of long-term flight but carrying only cameras rather than weapons or a payload—steadily began to move into the realm of domestic surveillance through loan-a-drone programs, in which the Department of Homeland Security provided drones to local law enforcement entities.[1] And in recent years as they've become cheaper and smaller, drones have become a tool for thousands of police departments.[2]

While action on drones has been severely lacking at the federal level, a significant number of states have been more proactive in establishing rules for how the government can use drones as a surveillance tool. Dozens of states have adopted rules and limits on drones, with 18 establishing a warrant requirement for police use of drones.[3] But numerous states have also passed reactionary measures prohibiting localities from passing rules and ordinances on drones;[4] although these state measures have some logical basis in preserving clear rules and consistency for small aerial devices, they prevent cities from advancing the privacy rights of their citizens. As more and more states take action, the desire for uniform, nationwide benchmarks will grow.

As mentioned, a successful law to limit drone surveillance has four main planks: a probable cause warrant requirement, an exhaustion requirement, minimization rules, and a requirement that the government provide logistical information with a justification for the extent of surveillance requested. Each plank is examined subsequently. In addition to these substantive limits, three major procedural factors that will help ensure success are worth considering: First, reasonable exceptions are necessary to deal with emergencies and other exigent circumstances. Second, special attention must be given to definitions in view of evolving technologies that could take advantage of loopholes. Third, sunsets should be in place that require lawmakers to regularly revisit policies for drones, both to respond to potential loopholes and to reevaluate what rules may be needed to respond to evolving technologies that could grow significantly more powerful. These procedural considerations are likewise examined.

REQUIRE PROBABLE CAUSE WARRANTS

The most fundamental and important component of an effective set of rules limiting drone surveillance is a probable cause warrant requirement for police drone surveillance. Independent review and authorization for surveillance based on suspicion of a particularized target is the foundation for almost all serious surveillance activities in the United States.[5] This standard began with the Founders' objection to British use of general warrants and desire to block the surveillance of large groups of people not connected to any specific wrongdoing. But throughout American history the development of new surveillance technologies has largely been followed (albeit often far too slowly) by the establishment of laws requiring independent judicial authorization for those activities—and requiring that surveillance be directed at a particular target.

This is not simply a long-standing tradition; independent judicial authorization accomplishes several goals that are fundamental to democratic government. First, it ensures invasive search activities are based on genuine suspicion of wrongdoing, limiting abuse and misconduct. Second, it prevents broad fishing for crimes and information that could lead to selective prosecutions. Third, it reasonably limits the scale at which government can search for and stockpile sensitive information about citizens. In the landmark *United States v. Jones* decision, Justice Sonia Sotomayor warned that government acquisition of "such a substantial quantum of intimate information about any person whom the Government, in its unfettered discretion, chooses to track . . . may alter the relationship between citizen and government in a way that is inimical to democratic society."[6]

Although judicial authorization based on suspicion of a target is a consistent theme, there is some variance in the level of suspicion required. For example, laws requiring judicial authorization for the government to obtain call records only require that the government show clear and material facts to establish that the records are relevant to an investigation, according to the concept that such material is less sensitive than the fruits of searching a house or wiretapping a phone call. Other forms of surveillance must be based on reasonable suspicion, an intermediate level of suspicion less than probable cause but more than mere relevance.

Therefore, the main question to consider in creating proper rules for drone surveillance is what level of suspicion should be required for law enforcement to use drones. There are several reasons why drones should be subject to a probable cause requirement.

First, tracking an individual's movements or monitoring an individual's presence at a specific location through electronic means that cannot be replicated by human labor is highly invasive.[7] As the Supreme Court held in its 2018 *Carpenter* decision, which dramatically expanded privacy rights by establishing a warrant requirement for cellphone tracking, privacy rights cannot be sustained if "With just the click of a button, the Government can access each carrier's deep repository of historical location information at practically no expense."[8] The precedent set in *Carpenter* is not only relevant for explaining why, as a principle, location tracking is so important that it necessitates use of probable cause warrants. It also set out a clear standard for the most commonly used means of location tracking. Similar types of tracking must be held to the same standard; not doing so creates a loophole.

Even if we view drones as too disanalogous to be placed in the same category as cellphone location tracking, the goals we strive toward in establishing judicial authorization all point to drones being a tool with significant risks, and in need of the strongest checks.

In terms of the need to prevent abuse, drones present huge risks. Drones can target sensitive activities, using overhead surveillance to monitor entire crowds. We've already seen aerial surveillance used for this type of pernicious targeting during protests: The FBI conducted large-scale aerial surveillance of thousands of demonstrators in Baltimore protesting police misconduct following the death of Freddie Gray in custody (this surveillance was conducted via manned aircraft but could just as easily have occurred using drones).[9] In Illinois, state lawmakers introduced a bill in 2018 to authorize use of police drones to monitor protests, and that bill was only defeated after the vocal objections of civil rights and civil liberties activists.[10] And there are a host of other activities that drones could target, with aerial surveillance providing a unique vantage point. Drones could monitor attendance at houses of worship, doctors' and lawyers' offices, or political organizations. This type of surveillance could be used for cataloging attendees; even absent misconduct, it could create a severe chilling effect that deters participation.

Absent strong judicial authorization requirements, drones could be used to gather information for intrusive fishing expeditions. Drones can continuously monitor huge areas; with some cameras a single drone's recording can encompass an entire city.[11] They could be used for automated flagging of petty offenses such as speeding or jaywalking. Or they could mark anyone engaging in "suspicious" activity—such as running down a city street—for further police action. This level of monitoring absent human engagement is worrisome in general, but especially problematic if applied in an improper manner. If law enforcement selectively deployed drones above particular communities—such as low-income neighborhoods or neighborhoods predominately populated by people of color—and programmed them to conduct this type of automated flagging, it could facilitate and give cover to discriminatory targeting on a mass scale.

Additionally, drones present a risk of generalized surveillance on a mass scale. Drones already have the potential to monitor huge areas and are a reasonably cheap purchase for local police departments: 500 drones designed for law enforcement could be bought for the same price as a single police helicopter.[12] Right now the biggest constraint on the government's ability to simply blanket an entire populated area with continuous drone surveillance is battery life; smaller, more affordable drones cannot run for significant lengths of time. But battery life is likely to improve over time; or alternatively, drones' overall cost could continue to decrease to the point that police departments can afford to maintain a small fleet of drones and run them in shifts in order to maintain continuous surveillance.

The potential for drones to continuously record aerial footage for most or even all of a populated area presents even more risk when combined with video analytic tools. Face recognition and license plate readers could allow for identification of individuals. The aerial surveillance company Persistent Surveillance Systems has developed an alternate means of identification: following individuals from a point of interest (such as a public event) to their home, and using that data point to infer identity. A host of other video analytic tools are available to law enforcement that could allow the government to quickly sift through huge amounts of aerial footage and find specific things or individuals. As Jay Stanley, senior policy analyst at the American Civil Liberties Union, notes in a report on emerging video analytic tools:

> Analyzing video is going to become just as cheap as collecting it. . . . While no company or government agency will hire the armies of expensive and distractible humans that would be required to monitor all the video now being collected, AI agents—which are cheap

> and scalable—will be available to perform the same tasks. And that will usher in something entirely new in the history of humanity: a society where everyone's public movements and behavior are subject to constant and comprehensive evaluation and judgment by agents of authority—in short, a society where everyone is watched.[13]

REQUIRE EXHAUSTION

Exhaustion is a requirement developed for invasive surveillance methods, most prominently wiretaps. The goal of "exhaustion" is to limit collateral damage to privacy by reducing situations where surveillance techniques that impact innocent individuals are employed.

In addition to probable cause requirements, the Wiretap Act stipulates that, to receive authority to conduct a wiretap, law enforcement must "exhaust" less invasive options, demonstrating to a court that traditional investigative techniques (i.e., those not involving electronic surveillance) have been tried and failed, or that they "reasonably appear to be unlikely to succeed if tried or to be too dangerous."[14] There are two main reasons for creating an exhaustion requirement for wiretaps. First is the severity of surveillance. Wiretaps record a host of sensitive communications, including many not likely to relate to criminal wrongdoing. Second, and perhaps more important, wiretaps by nature affect nontargets who are calling or receiving calls from the target. Even if someone believes that suspects should forfeit the privacy of all their communications by virtue of being a suspect (a problematic belief for reasons to be discussed later in the "Require Minimization Rules" section), it seems impossible to argue that other individuals should be treated as guilty by association and lose their privacy for this reason. A wiretap of a suspected mobster could pick up a call of a child discussing mental health issues just as easily as a conspirator discussing a criminal enterprise. Because they sweep in people beyond the target, wiretaps inherently cause collateral damage to privacy and create the need to exhaust less invasive methods first.

Drone surveillance by its nature engages in incidental collection of nontargets, recording video of everyone within the area of the person being monitored. If the target of drone surveillance goes into or near a sensitive location—such as a medical clinic, protest, or house of worship—all other individuals going to or from those locations are recorded as well. Before this type of surveillance with collateral damage to privacy is permitted, police should be required to exhaust traditional investigative techniques.

But the growing range of modern electronic surveillance techniques requires more than just the exhaustion of normal investigative procedures, as the Wiretap Act discusses. For location tracking alone, a huge array of new surveillance tools and techniques is already available for law enforcement to deploy. Some—such as cellphone location tracking—only gather information on the target and do not damage the privacy of other individuals. Other emerging surveillance techniques—including the use of cell-site simulators (commonly called "Stingrays"), collecting tower dumps that reveal all phones near a cellphone tower, and the use of drones—will always collect location data on everyone within a targeted area.

Given this range of options, and given that different options have a varying impact on innocent individuals' privacy, we should require a new form of electronic exhaustion.[15] Police who wish to deploy invasive electronic surveillance should not only be required to show that traditional investigative procedures have failed or would fail. They should also be required to show that the electronic surveillance technique being requested is not more invasive—particularly in terms of affecting nontargets—than other electronic surveillance techniques that could capture the necessary information.

So, for example, if police wanted to use a drone to monitor an individual's movements for a 12-hour period, they should be required to explain to a judge why a technique that is less invasive to nontargets, such as cell site tracking, would be unsuccessful. There are certainly situations where this would be the case: a surveillance target may exhibit a pattern of not carrying a phone when conducting illicit activities, or cell site tracking might provide insufficient precision. But surveillance techniques like drones that cause collateral damage to privacy should be the last resort, not the norm.

Another benefit of an exhaustion requirement is that it would make judges approving warrants better informed of the relative harm to privacy of different electronic surveillance tools and techniques. Individuals who are not deeply immersed in the new technologies, including judges, may lack a full understanding of how surveillance tools like drones and Stingrays can be limited. An added explanation of their impact, especially in relation to other available techniques, will have significant benefits in aiding judges' deliberations.

REQUIRE MINIMIZATION RULES

Minimization rules are another important tool to keep irrelevant private information from being swept up in electronic surveillance. "Minimization" requires the government to obscure information obtained during surveillance that is not necessary for that investigation. This can be achieved through full deletion or by replacing unnecessary details with more generic information (such as replacing someone's name with "PERSON 1" in a transcript). If law enforcement recorded a phone call or collected an email that was totally irrelevant to an investigation, minimization rules would typically require it be deleted. If intermittent pieces of a conversation were irrelevant, minimization would usually dictate replacing details with more generic information so that investigators could understand the evidence without unnecessarily gleaning personal information with no bearing on a case.

Minimization provides added privacy for nontargets swept up in surveillance as well as targets themselves. When collateral damage to privacy does occur, minimization limits the damage by removing the irrelevant information of targets and nontargets alike. This rule is essential because unrestricted monitoring of targets can lead to abuse. The most prominent example involved the wiretapping of Martin Luther King Jr. While the surveillance ostensibly occurred in response to suspicion that King had ties to communist agents, the information then FBI director J. Edgar Hoover and the FBI were most interested in was King's extramarital activities.[16] Hoover attempted to use this information to blackmail King and encourage him to commit suicide, one of the most infamous examples of surveillance abuse in U.S. history.[17]

Misuse of personal information swept up in the course of surveillance goes far beyond this one case. The Church Committee—a select committee led by Sen. Frank Church (D-ID) that worked in the 1970s to uncover government misconduct—found a myriad of cases in which the FBI's counterintelligence program (commonly known as COINTELPRO) took personal information completely disconnected from illegal activities and used it to target marginalized communities and dissidents:

> Groups and individuals have been harassed and disrupted because of their political views and their lifestyles. . . . Unsavory and vicious tactics have been employed—including anonymous attempts to break up marriages, disrupt meetings, ostracize persons from their professions, and provoke target groups into rivalries that might result in deaths. Intelligence agencies have served the political and personal objectives of presidents and other high officials. . . . Sometimes the harm was readily apparent—destruction of marriages, loss of friends or jobs. . . . But the most basic harm was to the values of privacy and freedom which our Constitution seeks to protect and which intelligence activity infringed on a broad scale.[18]

Given that drone surveillance is likely to capture the activities of a significant number of individuals whenever it occurs, this type of surveillance should include minimization requirements. The most basic minimization rule is an overall retention limit: if law enforcement has not designated collected data as evidence of criminal activities for a particular investigation within a set period of time, those data should be deleted. Additionally, law enforcement should be required to undertake efforts to obscure irrelevant details in footage that is otherwise useful. So, for example, if a drone followed a suspect in commission of criminal activities, minimization rules should require that individuals in other parts of the video not interacting with the suspect be blurred out.[19] These rules will prevent mission creep from occurring in the use of drones that leads to generalized surveillance of entire populations. Application of minimization should be especially strict when applied to surveillance of sensitive locations and activities, such as those where participation is protected by the First Amendment.

When minimization is required, the rules by which it occurs (date of deletion, etc.) tend to be developed by the agency itself, rather than proscribed in full detail within legislation. This method can be an effective delegation of authority, given that investigators likely have insight into how long it typically takes to evaluate particular types of potential evidence. But independent oversight is also needed. At a minimum legislation should set out general categories of minimization that should occur, including but not limited to general retention limits, obscuring individuals not relevant to footage used in investigations, and augmented protections for footage involving activities protected by the First Amendment. Legislation should also require that minimization rules be made public so that the legislature can act in a remedial manner if those rules are insufficient and require more direct requirements by statute. Regular auditing and independent oversight should occur to ensure that minimization rules are properly followed.

REQUIRE DESCRIPTION OF LOGISTICAL INFORMATION AND JUSTIFICATION FOR ITS EXTENT

In addition to the exhaustion requirement, the Wiretap Act sets out additional conditions on logistical details that can serve as an effective model for drone surveillance. The law requires law enforcement to supply numerous logistical details to explain the extent to which wiretapping will occur. Specifically, law enforcement must detail (a) the identity of the target; (b) the nature and location of communications facilities that will be tapped; (c) "a particular description of the type of communication sought to be intercepted, and a statement of the particular offense to which it relates"; (d) who is authorized to oversee the interception, and (e) the period of time for which the wiretap shall be active.[20] These factors help prevent unnecessarily broad surveillance that could lead to fishing and abuse, and they give the judge authorizing the wiretap a clear picture of the degree to which it will impact privacy.

Similar logistical details should be provided for drones. An application for drone surveillance should include the following logistical details:

1. the identity of the target;
2. the size and nature of the expected area to be monitored;

3. the period of time for which drone surveillance is sought;
4. who is authorized to operate the drone as it is recording and who is authorized to access footage; and
5. the type of activities law enforcement seeks to monitor using drone surveillance and what criminal offenses these activities relate to.

These details will give a judge a clear picture of the extent to which officers intend to use drone surveillance. A drone focused on one city block is far different from one that is monitoring a square mile. A drone monitoring a meeting between two individuals in an isolated field is different from a drone monitoring a populated downtown area. And a request to operate a drone for 30 minutes is different from a request to operate a drone for 24 hours.

Beyond simply providing this logistical information, investigators should be required to justify it. According to the type of activities investigators seek to monitor, they should explain why the expected area of surveillance is necessary to achieve their goal; if investigators want to monitor a 20-block area, they should explain why a 10-block area would be insufficient; and if they request use for a six-hour period, they should explain why five hours or three hours or one hour would be insufficient. These constraints will narrow surveillance to the best degree possible in a manner that is cognizant of investigative needs.

In addition to these four main rules designed to check and limit drone surveillance, legislators should keep in mind three other key factors: how to prepare for needs arising in emergency situations, the importance of forward-thinking definitions, and sunsets that require lawmakers to regularly revisit this issue.

EMERGENCY EXCEPTIONS

Rules governing all forms of invasive surveillance, from searches of homes to wiretaps, contain a series of exemptions for exigent circumstances. These include an emergency exception when action is necessary to prevent an imminent threat of death or serious bodily harm, and in some circumstances when action normally prohibited by law is necessary to continue a "hot pursuit" of a fleeing suspect. These exceptions should certainly apply to drone surveillance as well. Failure to include them would endanger public safety and undercut support for reasonable legislation limiting drone surveillance.

It is critical that the emergency exception be clearly defined and reasonably narrow. Including a carefully curated list of commonly used exceptions for surveillance with judicial authorizations should be a relatively simple task for legislators.

Because the two functions of drones—flying and filming—can be separated, one issue in this area is unique to drones: deciding exactly when exceptions—and the general limit on use of drones—should apply. It may seem natural to limit drone flights to situations in which a warrant has been obtained (and in fact, states that have a warrant-for-drone law typically regulate them this way). A better method may be to generally limit the authorization to *record footage* from airborne drones to situations in which officers have obtained judicial authorization.

Law enforcement may tout the benefits of aerial surveillance, highlighting the drone's ability to track a suspect in the moments after a crime has occurred; the drone can maintain pursuit while officers on the ground might not be able to.[21] But even with "hot pursuit" exigent circumstances removing the need for judicial authorization, launching a drone and tracking an individual fleeing a crime scene wouldn't be feasible in the time following the report of a crime. This scenario illustrates how attaching a warrant requirement (and the standard exceptions that accompany it) to the act of *recording* rather than the flight itself could provide dividends to police without infringing on privacy rights. If drones were allowed to remain airborne more generally, but only record on the basis of either obtaining a warrant or standard warrant exceptions, then an already in-the-air drone could respond rapidly to a crime in progress and conduct a hot pursuit in exigent circumstances.

This type of balance could enable a law with significant restrictions on when and how drones may be used to still play a role in situations where law enforcement believes drones will provide a major, unique benefit. Lawmakers and stakeholders need to strive to find mechanisms such as this one to resolve seemingly conflicting priorities, rather than hold them in inextricable opposition. Seeking ways to minimize disagreements is key to ensuring that more privacy-focused legislation is passed into law both at the federal level and in legislatures across the country.

DEFINITIONS THAT CANNOT BE EXPLOITED

A cliché for drafting legislation is that the devil is in the details and that lawmakers must write terminology extremely carefully. Even so, definitions for drones present a unique challenge. Because the technology is still relatively new, terms related to drones often do not have any colloquial meaning and need to be clearly defined. Most legislatures that limit drone surveillance define drones as vehicles that (a) fly, (b) are not operated by an in-vehicle pilot, and (c) use powered controls to remain airborne and control their movements. This format allowed Miami police to develop a loophole to the state's prohibition on drone surveillance when they created a system of tethering a blimp with surveillance cameras to a car.[22] Because the device—called a tethered aerostat—used buoyancy rather than a motor to remain airborne, and because it used the tethered vehicle on the ground to control movement, it did not fall within the state's definition of "drone." City Manager Jimmy Morales stated that the explicit goal of purchasing the surveillance blimp was to acquire an aerial surveillance device that would get around the Florida state ban on drone surveillance.[23]

In another example that is less a loophole than an alternate system with similar impact, police have turned to manned aircraft for large-scale aerial surveillance. The Persistent Surveillance Systems program that Baltimore police deployed for months used manned Cessna planes and high-powered cameras to create an incredibly powerful aerial surveillance system.[24]

Both of these situations are different from cases that involve drones. A tethered aerostat is somewhat limited in its movement and far less agile than a drone. The Persistent Surveillance System requires manpower for the flights and is significantly more expensive than drones. Lawmakers may wish to treat each technology as a distinct item, draw one line between manned and unmanned aircraft, or have a single overarching policy on aerial surveillance. All options have merits and drawbacks. Lawmakers developing legislation on drones and aerial surveillance should carefully consider how broadly they want their policy to apply and then carefully draft definitions according to that goal.

SUNSETS IN DRONE SURVEILLANCE LEGISLATION

Laws authorizing new and powerful surveillance techniques sometimes include sunset provisions that force lawmakers to revisit the issues and reevaluate the laws' impacts on a regular basis. These sunset provisions can be highly valuable in forcing consideration of surveillance laws that may have unintended consequences, may be controversial in their definitions or how broad the authority is, or may lack transparency about what type of surveillance is occurring and what, if any, public benefit it provides.[25] All of these apply to drone surveillance.

Additionally, manufacturers continue to rapidly improve drones' capabilities.[26] Already powerful cameras attached to drones are likely to continue to advance. Drones themselves are likely to move faster, maintain higher attitudes, and—most important—stay airborne for longer periods of time. Automation functions, such as those allowing drones to "lock onto" and follow a specific target without a human operator, and swarm functions that allow a single pilot to control multiple drones, will become more common and more effective. And the overall cost of drones will continue to decrease, allowing police departments to deploy more and more drones to cover increasingly large areas.

In 5 or 10 years, the capabilities of drone surveillance will likely be significantly different from what they are today. It would be an enormous mistake to think that right now we can effectively anticipate and account for how invasive drone surveillance will be in the future. Even with rigorous limits in place, technological advances over the next several years may necessitate far more restrictive rules—or a full prohibition—on drones and aerial surveillance.

Therefore, any statute on law enforcement use of drones should contain a sunset provision: after a set period of time, lawmakers must revisit the laws regarding drone surveillance and either reauthorize them, or let the authority for law enforcement to use drones expire entirely.

CHAPTER SEVEN

DRONE CAPABILITIES AND THEIR USES BY THE FEDERAL GOVERNMENT

JAY STANLEY

SURVEILLANCE CAPABILITIES OF DRONES

Drones, or unmanned aerial vehicles, have been with us in various forms for a long time, but in the past decade they have become an increasingly common presence in American life.[1] Today their integration into daily life is poised to reach a whole new level as their technological capabilities and legal latitude for operation expand. That increased presence will likely bring certain conveniences and efficiencies but will also make drones an increasingly powerful tool for surveillance. As that happens, they will become a tool of growing interest to local, state, and federal law enforcement—and the importance of understanding just what the capabilities of these machines are will likewise grow.

Privacy advocates have been concerned about modern drones since at least the 1990s.[2] The decade that followed saw gradually increasing discussion of the technology as a military tool, combined with science fiction–like speculation about their dangers for domestic privacy. Then, in December 2011, the *Los Angeles Times* published a story about a rural sheriff who had a warrant to search a farm for six missing cows and requested that U.S. Customs and Border Protection (CBP) use one of its Predator B drones to help him carry out his search.[3] With that story, which received widespread public attention, domestic drones began to rocket toward the top of public attention and blossomed into a full-fledged political issue. Within a year, a wave of legislation had been introduced around the country regulating the domestic use of drones; by mid-2014 such bills had been introduced in 36 states and enacted in 13.[4]

The deployment of drones stalled, however—held back by the Federal Aviation Administration (FAA), which imposed significant restrictions on drone flights, including a 400-foot altitude limit and bans on night flight, flights over people, and flights beyond the visual line of sight (BVLOS).

The pieces are starting to finally come together for more deployments, however. The FAA is slowly building a legal infrastructure for significant increases in drone usage in American life. For example, in 2019, the agency began a rulemaking process to ease the current ban on nighttime operations and flights over people.[5]

The biggest remaining obstacle to expanded drone flights, however, is the FAA's rule against BVLOS flight. Although the agency has been issuing special exemptions from the ban to a handful of test sites and other operators since 2015, routine BVLOS flight is still forbidden.[6] A big stumbling block is security. A regulatory framework permitting routine BVLOS flights is not going to happen, as an FAA advisory committee of outside experts put it, "until the law enforcement and national security communities are comfortable with their ability to identify and track [unmanned aircraft]."[7] As a result, in December 2019, the FAA proposed a national system for the real-time tracking of drone flights.[8]

Other obstacles are technological. The FAA has said that it won't allow routine BVLOS flights until unmanned aircraft gain a reliable ability to detect and avoid other aircraft—especially with regard to large drones. Much civilian aviation has long been based on "see and avoid"—basically, the requirement that pilots look around and make sure they don't collide with other aircraft. Conferring that ability on machines, however, is not a simple task; it depends on reaching a reliable level of artificial intelligence (AI) machine vision or other spatial detection abilities. Other technology challenges include ensuring that the communications link between drones and ground controllers can be made reliable (including what happens when the dreaded "lost link" takes place) and figuring out how airspace is best managed to ensure that drones don't collide with each other or with manned aircraft.[9]

Ultimately, the FAA's vision is this: unmanned aircraft systems (UAS) "operating harmoniously, side-by-side with manned aircraft, occupying the same airspace and using many of the same air traffic management (ATM) systems and procedures."[10]

As that happens, the surveillance capabilities that those drones can carry will become increasingly significant—and increasingly important for the public and policymakers to be aware of. Drones and the surveillance devices they carry are a fast-moving area of technology, and assessing their capabilities requires us to have one foot in the present and one in the near future. Technology development is nonlinear and cannot be easily predicted, but we can make reasonable guesses in many areas.

DRONES AS A PLATFORM

Drones themselves are not capable of any surveillance. They are a *platform* on which operators can attach surveillance equipment. The only limits on what such equipment could be are what has been invented, what is practical and desirable to use in the air, and size and weight. Among the sensors that can be attached to drones are GPS, radar, lidar, rangefinders, magnetic field change sensors, sonar, radio frequency sensors, and chemical and biochemical sensors. And of course, cameras are probably the most significant surveillance equipment that drones put into the air. Many cameras include thermal and other sensors that collect signals beyond the visual part of the electromagnetic spectrum.

Drones could also be used to carry equipment for electronic signals collection. For example, law enforcement has in the past decade begun using devices known as cell-site simulators (popularly known as "Stingrays") that are essentially fake cellphone towers. Like real towers they broadcast a signal that prompts any mobile phone in range to identify itself to the device. They can thus be used to collect the identities of people in a particular area en masse. The FBI as well as other law enforcement agencies such as the U.S. Marshals have acknowledged that they sometimes use an aerial version of these devices, known informally as "dirtboxes," on manned aircraft.[11] We don't know of any deployments on drones, but as extended drone flights become increasingly possible, such deployments are very likely.

CAMERA TECHNOLOGY

In August 2016, *Bloomberg Businessweek* revealed the existence of a pilot program being operated by the Baltimore Police Department in which small manned aircraft circled over the city all day, using megapixel cameras to continuously photograph a 32-square-mile area and giving police the ability to retroactively track any vehicle or pedestrian within that area.[12] It was the ultimate Big Brother "eye in the sky"—and yet the Baltimore police had not notified the public or even the mayor or city council about the program. Revelation of the secret pilot program generated a storm of controversy, and eventually it was put on hold—though in December 2019, the city's police commissioner announced that the program would be revived.[13]

The Baltimore program and brief tests in other cities before it have been run by a private company called Persistent Surveillance Systems (PSS). The company's surveillance is currently carried out using manned aircraft. But if Americans decide to allow this kind of ongoing aerial surveillance over their communities, drones will almost certainly replace manned aircraft, if and when the FAA permits the kind of flights involved.

The PSS technology is often referred to as wide-area aerial surveillance or wide-area motion imagery (WAMI). It is based on a military predecessor called Gorgon Stare, a wide-area surveillance device that was developed by the military and first deployed in 2012 (with a second, more powerful version unveiled in 2014).[14] Affixed to Reaper drones, Gorgon Stare was used in overseas theaters such as Afghanistan and Iraq. Other companies also advertise wide-area surveillance capabilities for domestic use, though there has been no public word of any domestic uses beyond testing by those companies.[15]

The technology behind all of those programs is similar: it points multiple cameras toward the ground and stitches those images together into a single, larger photograph. It also uses computers to automatically correct for the changing camera angles of the circling planes as well as factors such as topographic variances and lens distortion.[16]

The result is a surveillance system of enormous power, able to reconstruct the movements of all visible vehicles and pedestrians across a city—where they start and finish each journey and the paths they take in between. That can allow tracking of a great proportion of people's movements throughout a city.

There are limitations to aerial surveillance by drones or any other platform: surveillance can be impeded by darkness, cloud cover, and subjects entering tunnels and buildings; views can also be obstructed by tall buildings, foliage, and other obstacles. In a downtown area with many tall buildings, many subjects will not be visible except when a camera is directly overhead, which in some circumstances will be only a small proportion of the time. Most vendors offer infrared cameras for night surveillance, and other sensor technologies such as lidar can be used in some circumstances to see through clouds and tree cover.[17] But those technologies have limitations, including resolution and clarity when compared to light-of-day photography.

Another determinant of a drone camera's effectiveness for surveillance is the camera's image quality. Most surveillance cameras in common use today have resolutions of 1080p (1920x1080 pixels), 4 MP (2688x1520), or 4K (3840x2160)—although cameras with equal pixel resolutions can vary greatly in image quality due to numerous factors, including lens quality, compression level and technique, frame rate, shutter speed, dynamic range, and low-light performance.[18] Among the complexities is the fact, surprising to many, that lower-resolution cameras often have better low-light performance because they have larger pixels able to gather more light.

Overall, however, the image sensors on surveillance cameras are steadily becoming more powerful.[19] Ultrahigh-resolution cameras are already available that can capture hundreds or thousands of megapixels—"so much detail [that] you can pick out a face amongst a crowd of thousands," as one vendor brags.[20] Canon says that its 250-megapixel sensor can "read the lettering on the side of an aircraft flying 11 miles . . .

away."[21] The ultrahigh-resolution imaging system on the military's original Gorgon Stare camera, as reported by PBS *NOVA* in 2013,[22] used 368 smartphone sensors of 5 megapixels each, which were combined into a single 1.8-gigapixel camera that could cover 36 square miles at a resolution of 0.15 meters ground sample distance (GSD, the distance on the ground from the center of one pixel in a photograph to the center of an adjacent pixel). PSS, meanwhile, says that its camera features 12 16-megapixel cameras that produce 192-megapixel images covering 32 square miles at a resolution of 0.5 meters GSD.

Other companies are experimenting with composing even larger photographs using this technique of stitching together smaller ones. One company, for example, made a 57-gigapixel photograph as part of an advertising campaign.[23] Gimmick or not—and it required a painstaking effort to make—this technique will likely become easier and more powerful over time. The Department of Homeland Security (DHS) has already created a 360-degree, 240-megapixel spherical camera made up of 48 smaller cameras "to achieve wide-area situational awareness" in places such as sports stadiums—and has already deployed it in one: Seattle's CenturyLink Field.[24]

Even as the images are getting bigger, the equipment necessary to take these photographs is getting smaller. Wide-area surveillance cameras, which once required larger aircraft, can now be installed on small tactical drones weighing less than 150 pounds.[25]

In addition to image resolution, surveillance imagery is also measured in temporal resolution. Most standard U.S. video cameras take 30 or 60 frames per second, while most industry-grade surveillance cameras capture 8–10 frames per second to reduce storage costs. Gorgon Stare reportedly captures 12 frames per second, while PSS says its cameras capture just 1 frame per second. That low frame rate reduces data bandwidth and storage needs, and yet still allows people and vehicles to be effectively tracked in most circumstances. In theory that low frame rate could cause problems for surveillance in some circumstances—some people can easily run 15 feet or more per second.

Technologists are pushing the limits of camera technology in other ways too. In recent years, thermal (infrared) cameras have become both smaller and less expensive even as their resolution and image quality have improved.[26] Deep learning is also being used to fuse infrared and visual data into "multispectral images" in which optical and infrared images are combined to bring out details that neither can pick up alone, helping overcome problems such as low light and distant subjects.[27] AI deep learning techniques are also being used to improve low-light photography.[28]

AI techniques are also being used with increasing success to virtually "up-size" images beyond the resolution at which they were filmed. Limited versions of this technique can sometimes be made to work because multiple lower-resolution images with slightly different perspectives contain more information about a scene than is contained in any one image.[29] Other AI techniques use models of the real world to guess at the content of missing pixels.[30]

AI VIDEO ANALYTICS

For all the ways that the imagery captured by drones has improved and is likely to do so further, some of the most significant advances in aerial surveillance are and will be happening not in the creation of the underlying photography, but in the software that is used to analyze all those data.

When PSS, the company running the Baltimore surveillance system, gets a report of an incident from the police, it uses human analysts to call up the images from the relevant time period and manually trace the movements of visible pedestrians and vehicles backward and forward in time as necessary for the investigation. Many people are interested in automating that process, however. Driven by startling advances in machine learning and AI, a vast field of "video analytics" has emerged that aims to teach computers to "watch" and interpret video so that humans, who are expensive, slow, easily distracted, and difficult to scale, do not have to.[31] Researchers are training computers to recognize different human actions, behaviors, and characteristics in surveillance camera footage—and to understand the all-important contexts in which those things are observed. Although progress is slow and obscured by clouds of hype and exaggeration, such automated visual monitoring may become one of the most significant surveillance methods yet developed.

Researchers are also doing a lot of work specifically in applying video analytics to aerial imagery. Aerial imagery analysis is difficult, experts say, posing many challenges not present with fixed, ground-level cameras. These challenges include "small and low resolution targets, large moving object displacement due to low frame rate, congestion and occlusions, motion blur and parallax, camera vibration, camera exposure and varying viewpoints in addition to background variance, illumination changes or shadow interference."[32]

Researchers have compiled collections of aerial video as training data sets to help AI programmers teach computers to overcome such challenges and recognize objects, actions, and movements in aerial images.[33] The military has also released an enormous collection of aerial imagery for AI training and is offering a reward for those who can best analyze it.[34]

A major focus among researchers is creating the ability to track "tagged individuals or vehicles" across a city—automating the process that PSS's human analysts currently carry out manually. Researchers are also aiming to train computers to track a large number of targets at once.[35] The military, as the primary institution that has deployed wide-area surveillance so far, has been working on this for more than a decade—currently through an effort dubbed Project Maven.[36] Maven aims to use object recognition to automatically recognize vehicles and as many as 38 other categories of objects filmed by drones, and to track them as they move about. As one of the Pentagon's first forays into using machine learning, this project received enormous media attention in late 2018 when protests by Google employees pushed that company to cease work on it.[37] (The contract was subsequently picked up by the company Palantir.[38])

Recording the movements of everyone in a city represents an enormous and new surveillance capability, equivalent to GPS tracking an entire population. As the U.S. Court of Appeals for the D.C. Circuit explained in a GPS tracking case:

> A person who knows all of another's travels can deduce whether he is a weekly church goer, a heavy drinker, a regular at the gym, an unfaithful husband, an outpatient receiving medical treatment, an associate of particular individuals or political groups—and not just one such fact about a person, but all such facts.[39]

And of course, different people's travels can be cross-referenced, providing detailed associational information as well—who spends time with whom.

PSS and its defenders protest that their system does not take sufficiently high-resolution photographs to directly identify vehicles or pedestrians (which is often difficult even with high-resolution overhead imagery because of the angle). However, the company's aerial images can easily be cross-referenced with time-stamped video from ground cameras or any other available source of data, such as tollbooth transponders or credit card purchase records, to identify subjects in those aerial images. Even without those ground sources of data, people can often be identified from their locations alone. Even relatively rough location information about a person will often identify them uniquely. For example, according to one study, just knowing the census tracts (roughly equivalent to zip code) of where a person works and where he or she lives will uniquely identify 5 percent of the population. Using the "census blocks" where somebody works and lives (an area roughly the size of a block in a city, but larger in rural areas) yields an even higher accuracy level, with at least half the population being uniquely identified.[40] And often WAMI will permit more precise data than that by revealing a vehicle or pedestrian's home address. Academic papers have been written about inferring home address from location data sets.[41] Add a specific work address and that would most likely identify virtually anybody.

All this is possible just with manual analysis of aerial imagery. Using AI to actually extract all visible movement data from that imagery opens up even broader and deeper monitoring possibilities. Computer algorithms could create searchable stores of location data ("show me all the places this particular vehicle has been, and all the vehicles spotted at this particular location") as well as scrutinize and evaluate people's movement patterns in a search for "suspicious activity" or the like.

The military has engaged in just such efforts. For example, it trained algorithms to help with the detection and investigation of insurgent activity, such as the planting of roadside bombs in Iraq. One goal has been to carry out "pattern of life" analysis on subjects and communities—to detect regularities in their movements and activities to discover things about them, predict where they will be, or sound alerts when variations and anomalies in those patterns arise.[42] The military has also worked on "risk propagation": given a suspicious subject, the locations that the subject visits are recorded, and other people who also visit those locations are then also tagged as potentially suspicious.[43] Algorithms could also scrutinize citywide streams of imagery and location data for such things as heat maps and "change detection" (flagging changes in buildings, plants, parked vehicles, or other parts of images), and for setting "watchboxes" and "trip wires" (flagging vehicles that enter specified areas or cross specified lines).

Academic work by machine-vision specialists and other scientists suggests the depth of predictive and identifying analysis that could be done given the location data sets that automated analysis of WAMI could produce. One study found that most people are so routine in their movements that "our mobility is highly predictable at a city-scale level" such that "the location of a person at any given time can be predicted with an average accuracy of 93% supposing three km^2 of uncertainty."[44] Even for those who travel frequently the predictability level is 80 percent, according to another study, which concludes, "the development of accurate predictive models is a scientifically grounded possibility."[45] Some researchers are working to predict movement "at a finer resolution such as in shopping malls, in airports, or within train terminals," and propose an algorithm to "model the social interactions of pedestrians to predict their destination."[46]

DRONE FLIGHT BEHAVIOR

Another area of intense research when it comes to drones is autonomy. Most "unmanned" aerial vehicles today are not actually pilotless; they're just piloted remotely. But researchers are doing much work to teach drones to navigate and pilot themselves. Already some consumer drones contain semi-autonomous features such as "return to home" and "follow me" functions (in which drones automatically return to a defined home or follow a subject) and the ability to set predefined flight paths that the drone follows without further instruction.[47]

More ambitious work is being done by researchers on automated sense-and-avoid technology aimed at preventing drone collisions with objects including other drones. Some consumer drones already claim such capabilities, and research continues on a number of different sensing technologies, including cameras, ultrasonic sensors, lidar sensors, and microwave radar.[48] The development of reliable sense-and-avoid capability is critical for BVLOS flight, which is in turn critical if drones are to carry out functions such as deliveries—and wide-area surveillance—which analysts expect to become a major growth industry.[49]

Indeed, autonomous flight can mean more than navigating back to home and avoiding telephone wires; it can also be used to create swarm behavior in drones, in which a number of independent drones work cooperatively to carry out tasks that a single drone could not. The military has already begun experimenting with aircraft-launched drone swarms[50] and has put out a call to drone developers and AI experts to explore whether autonomous swarms could be useful in search and rescue operations.[51] It is not a stretch to foresee a police department implementing an inexpensive, high-resolution, wide-aerial surveillance system made up of a swarm of cheap, autonomous drones automatically spreading out to cover a region, taking turns flying "shifts" with other drones and docking themselves at charging stations in between. Researchers have already built prototypes of systems that stitch together video from multiple drones spread across an area into a single wide-area image.[52]

Another major goal of drone makers is to develop a craft that can stay aloft longer. In December 2019, the U.S. Air Force tested a craft it calls Ultra LEAP (for Long Endurance Aircraft Platform) with a successful two-and-a-half-day flight over Utah. The Air Force said that "subsequent test flights will demonstrate increased levels of flight endurance," and that such long flights would help solve the "tyranny of distance" problem for intelligence, surveillance, and reconnaissance tools.[53]

Balloons, of course, can stay aloft for much longer periods of time than rotor-blade or fixed-wing drones, and can make excellent platforms for surveillance equipment. They can also navigate, albeit without the precision of powered craft, by rising and dropping in altitude to catch winds moving in different directions.[54] Powered blimps and dirigibles can also have very long flying times. Blimps tethered to the ground, called aerostats, are also used for surveillance.[55]

"Drone" typically refers to fixed- or rotor-wing craft, but the definitional boundaries among all of these aerial craft are somewhat arbitrary; when it comes to function, any of them can be used for surveillance. Even the boundary between a drone and an aerostat is not always a bright one; like aerostats, some traditional drones contain tethers providing ongoing power and data connections to the craft while it's aloft.[56]

Solar-powered drones have become a major center of research and development investment in the past several years, with major companies such as Airbus and Google developing products in cooperation with NASA and the Defense Advanced Research Projects Agency, or DARPA. Airbus has already developed a spy plane that can stay in the air for more than two weeks.[57] Boeing is working on an aircraft that it says will be capable of remaining aloft for years at a time.[58]

Some of these craft are designed to fly at extremely high altitudes, where they are envisioned as remaining to provide surveillance, communications, or weather functions. There is an entire category of craft called "atmospheric satellites," so-called because they function as a sort of hybrid between space-based satellites and regular atmospheric aircraft. In 2019, one unmanned glider reached an altitude of 98,000 feet, well into the stratosphere.[59] So far, all the aircraft tested or developed in this category have been unmanned aerial vehicles (UAVs).[60]

ARMED DRONES

In July 2016, a gunman opened fire at police officers in Dallas, killing five and wounding seven others. The shooter was cornered in a parking garage, and police said that when negotiations "broke down" after a number of hours, the department repurposed a bomb-defusing ground robot with an explosive charge and drove it near the gunman before detonating the charge, killing the gunman. The incident kicked off a national discussion about police using "killer robots."[61] Under a 1985 Supreme Court case, *Tennessee v. Garner*, as well as other cases, the police may not constitutionally use deadly force unless someone represents an imminent threat to others and the use of lethal force is a reasonable last resort.[62] Efforts to arm domestic drones are widely (and wisely[63]) seen as beyond the pale, and for the most part lawmakers and law enforcement officials have not yet seriously contemplated using armed drones. The International Association of Chiefs of Police has recommended against arming UAVs, for example.[64] Nevertheless, no discussion of the capabilities of drones would be complete without acknowledging their potential to serve as a platform for weapons.

There are exceptions to the general taboo against arming drones. One sheriff in Texas mused about mounting less-lethal weapons like rubber bullets on drones.[65] The Electronic Frontier Foundation uncovered CBP documents suggesting possible future enhancements to its drone program, including "non-lethal weapons designed to immobilize [targets of interest]." CBP denied any plans to arm its drones with "weapons of any kind."[66] And in 2015, North Dakota enacted a drone bill that explicitly permitted law enforcement to equip drones with less-lethal weapons such as rubber bullets and tear gas.[67]

Arming drones is something that is entirely feasible, and there is good reason to think that at some point, if and when drones become a common part of everyday life, we will see further weapons deployments.

FEDERAL GOVERNMENT USES OF DOMESTIC DRONES

As the capabilities of drones increase—along with awareness of those capabilities, expertise in utilizing them, and the leeway that they are given in the national airspace—governmental interest in using them for surveillance and other purposes has grown and will no doubt continue to grow. As one former Pentagon official put it, "Officials outside of the military now have a better understanding of what drones are and what they can do, so it is not surprising to see requests for their use growing within the United States government."[68] Drones are used by all levels of government, but the remainder of this chapter will focus on federal uses, as local uses are covered elsewhere in this volume.

Like everybody else, U.S. federal agencies have been held back in their use of drones by the FAA's cautious approach toward the integration of

UAVs into the national airspace—as well as by shortcomings with the technology itself, such as drones' generally limited flight time. If, as anticipated, the federal government makes greater use of the technology over time, the deployments of drones that have been made so far are a good indicator of the directions they'll likely go in the future.

BORDER DRONES

The most prominent and consistent domestic use of drones by a federal government agency is CBP's use of drones at the nation's borders. The agency began flying drones along the southwestern border in 2006 and has since expanded its deployments to the Canadian border, the Caribbean, the Gulf of Mexico, and the southern California coast. Since 2011 the craft have been operated from three National Air Security Operations Centers located in Sierra Vista, Arizona; Corpus Christi, Texas; and Grand Forks, North Dakota. In 2017, according to the Department of Homeland Security (DHS) inspector general, the CBP completed 635 flights from its three operations centers.[69]

CBP uses MQ-9 Predator B ("Reaper") drones, as well as a variant called the "Guardian," which is used in maritime operations. These are large craft, with a wingspan of 66 feet and able to carry a payload of 3,850 pounds. They have a claimed ability to fly for up to 27 hours, a maximum speed of 276 miles per hour, and a maximum range of 1,878 miles.[70] Their maximum altitude is 50,000 feet, though the CBP missions fly no higher than 28,000 feet (they're required by the FAA to remain above 19,000 feet and no more than 60 miles from the southern border and 100 miles from the northern).[71] Each drone is operated by three employees: a pilot, a technician, and a "sensor operator" who monitors the surveillance feeds, zooming in on suspected targets where desired.[72]

These craft carry a range of video, radar, and other sensing technologies.[73] Prominent among them is a system called Vehicle and Dismount Exploitation Radar (VADER).[74] VADER is based on synthetic aperture radar, a technology that simulates a much larger antenna than would otherwise be practical by leveraging the fact that a moving aircraft covers a wide span of space as it moves. It can create much higher-resolution three-dimensional images than conventional beam-scanning radars, and because of advances in electronics the technology has become increasingly economical for small-scale uses.[75] Synthetic aperture radar can create high-resolution still images and a "real-time ground moving target indicator" that works "by detecting the Doppler shift that moving objects produce in radar return signals."[76] This, as one CBP official told a reporter, lets analysts see either still images that "look something like high-contrast black-and-white photos" or "moving targets displayed as dots superimposed on a map."[77] One source calls it a "man-hunting radar."[78] VADER is also used for change detection, with resolution that makes it "capable of seeing even small changes to a scene such as tire tracks and footprints." "We can take a picture, go back and take another picture later of the same thing and run a computer algorithm to determine what has changed in those two pictures," the CBP official said.[79]

In a 2013 document, CBP revealed that certain manned aircraft also carry "sensors used to detect electronic signals in the electromagnetic spectrum," working "in support of counter-terrorism efforts and to interdict organized smuggling." There's no reason to think such sensors could not also be attached to drones.[80]

It should be noted that the value and effectiveness of CBP's Predator program are highly questionable. Repeated reports from the Government Accountability Office (GAO) have found serious problems with the program, including very low actual availability of the aircraft for flight, shorter than advertised flight times, high susceptibility to disruption by clouds and other weather, and crashes. GAO findings suggest that the drones may have detected at most 1.2 percent of illegal border crossings, making them very expensive in proportion to their effectiveness.[81]

SMALL UAVS

In 2017, CBP began supplementing its large military Predator aircraft with a small unmanned aerial systems (sUAS) program, using drones weighing less than 55 pounds.[82] Those craft cannot fly as high or as long as the larger craft but are cheaper, are more efficient, and require less training and infrastructure. The best-performing craft the agency uses have a maximum speed of 50 miles per hour and an endurance of 60–90 minutes. The government says they carry "payloads such as video surveillance systems, rangefinders, thermal imaging devices, and radio frequency sensors."[83] They are also reportedly incorporating video analytics technology to "automatically sense if there is a suspected person where there shouldn't be."[84]

CBP says it uses the small drones to patrol the border, in investigations, for damage assessment, and for officer safety "in support of agents on the ground." Under the terms of its authorization with the FAA, the devices have to remain below 1,200 feet in altitude and are confined to "sparsely-populated locations."[85] CBP plans to continue expanding its sUAS program and in 2019 placed an order for around 100 more of the devices.[86]

The agency is also testing semi-autonomous small drones that can "launch from and land on the bed of a moving vehicle" and have "fully autonomous navigation" as well as "advanced computer vision capabilities" including "automatic target detection and geolocation."[87]

As such language indicates, CBP is definitely interested in exploiting emerging video analytics. "I don't think there's an organization on the planet that doesn't want to do something more efficiently using AI," Ari Schuler, a CBP technology director, told a reporter in 2019. "The art of letting a computer see like a person is tremendously valuable."[88]

WIDE-AREA SURVEILLANCE

One of the biggest questions is whether CBP has already or will in the future deploy wide-area surveillance systems. Arthur Holland Michel, former codirector of Bard College's Center for the Study of the Drone, believes that the agency is "likely to be the first U.S. federal agency to acquire WAMI for routine operations."[89] It's difficult to find information on how wide an area the VADER radar system sweeps, at what resolution, and with what kinds of analytics used to crunch the imagery that it collects. In a fact sheet advertising a version of the Predator B that has been "optimized for Intelligence, Surveillance, and Reconnaissance" called "SkyGuardian," the aircraft's manufacturer, General Atomics, describes its sensor package, which sounds very similar to VADER and may be what CBP is using. The package provides "broad-area stripmap coverage out to a range of more than 80 km, with a high-resolution spot capability down to an exceptionally fine resolution," and a Ground Moving Target Indicator mode that can "detect and track moving vehicles out to a range of 23 km."[90] Another source says that VADER's "exploitation suite" includes tools that enable the "persistent reconnaissance, surveillance, tracking, and targeting of evasive vehicles and people moving on foot in cluttered environments."[91]

We also know that CBP has solicited wide-area surveillance systems at the border, and in 2001 tested the use of the 440-megapixel Kestrel system (used for wide-area surveillance in Afghanistan and Iraq) attached to aerostats.[92] CBP still maintains a Tethered Aerostat Radar System program consisting of eight enormous aerostats hovering at an altitude of 10,000 feet along the southern border. The agency emphasizes their use in trying to detect low-flying aircraft trying to cross the border using radar, but also describes them as "the only persistent wide-area air, maritime and *land* surveillance system specifically designed for CBP's border security mission."[93] CBP is also testing smaller aerostats for other surveillance roles, including a model called the Persistent Threat Detection System, which has been deployed in the war theaters of Iraq and Afghanistan since 2003. This model has various sensors, including an "acoustic-sensor array that detects, locates," and directs a camera to focus on "transient sounds, such as enemy mortar, rocket launches, and IED attacks, and calculates the ground location of the threat source."[94]

CBP has also made investments in systems designed to process the oceans of data created by wide-area surveillance systems—a "strong indicator that it plans to buy into WAMI sooner or later," as Michel puts it.[95] In 2013, CBP also disclosed that it had deployed a drone on the southwestern border with something called the Wide Area Surveillance System (WASS):

> WASS uses a sensor mounted to the wing of a UAS to sweep large areas of border territory (approximately six kilometers in width) as the aircraft moves along its flight path. WASS alerts CBP to the existence of persons and/or vehicles along the border and provides coordinates to determine their location. The UAS pilot and sensor operator can then inform ground units of the location so that Border Patrol may coordinate an interdiction of the persons or vehicles. WASS provides a radar sensor image.[96]

Strangely, no further mention of this system appears to have been made in subsequent documents.

Although wide-area surveillance at the border would pose an enormous invasion of privacy for those who live in border regions, CBP likely regards the technology as a good way to augment its mission. The agency has shown little hesitancy in instituting measures that have marked negative effects on border communities, including oppressive internal checkpoints, ominous hovering blimps, and ground surveillance technologies such as surveillance cameras, license plate readers, and surveillance towers. Surveillance towers are proliferating rapidly in border communities; the agency says they "provide long-range persistent surveillance," "cover very large areas," and "automatically detect and track items of interest." The system sends "the data, video and geospatial location of selected items of interest" to a centralized data hub "to identify and classify them."[97]

And of course, it is important to remember that those kinds of ground surveillance systems can be, and in the future likely will be, integrated with drone surveillance. This can be accomplished in much the way that Baltimore police have used ground cameras to augment the low-resolution images created by PSS, creating surveillance synergies that are more powerful than the sum of their parts.

LITTLE CONCERN FOR PRIVACY

CBP as an agency does not appear to have much institutional history or knowledge when it comes to handling the privacy issues that arise when data are collected about American citizens. That may be because historically its role has been confined to the actual border where (at least until the advent of new technology[98]) the boundaries of travelers' privacy were well established by law and tradition. Privacy has thus been much less of a counterbalancing consideration than it is for the mission of internal law enforcement agencies. Thus, the agency failed to consider the privacy implications of its border surveillance technologies—and failed to comply with legally required privacy reviews. "Various CBP officials told us," the DHS inspector general wrote, that "they were unaware of the requirement to complete a [privacy review] before deploying" surveillance systems.[99]

Similarly, CBP has rushed to embrace face recognition, heedless of the enormous privacy implications—and controversy—that the technology carries. Not only is it pushing face recognition in airports, it also has sought to use that technology on drones. In 2017, the agency published a contractor solicitation notice seeking small drones that include "facial recognition capabilities that allow it [to] cross-reference any persons identified with relevant law enforcement databases."[100]

When Congress gave CBP the authority to fly drones at the border, it was undoubtedly to safeguard the border. But the agency has made a practice of regularly loaning its Predators to other law enforcement agencies inside the United States for a range of other purposes. These loans effectively turn CBP's border protection program into a legal Trojan horse, allowing entry of large, military-grade drones into American life in a way that has not been authorized. A records request by the Electronic Frontier Foundation showed that from 2010 to 2012, CBP flew nearly 700 missions for other agencies. Those included immigration agencies; the FBI; the Drug Enforcement Administration; the Bureau of Alcohol, Tobacco, Firearms and Explosives (ATF); the U.S. Marshals Service; and state police departments. The purposes ranged far afield from CBP's border patrol mission and included such things as investigating fishing violations and searching for marijuana plants, drug labs, and at least one case of missing cows.[101]

DOMESTIC MILITARY FLIGHTS

Besides CBP, the military is the only arm of government that is permitted to routinely fly large, high-altitude drones in the national airspace.

In August 2017, a mysterious, unmarked military aircraft was spotted flying circles over Seattle, Washington, for two weeks. Various military officials denied any knowledge of the craft, leading to much speculation about who was flying it, what it was doing, and what kind of surveillance equipment it might be carrying. Finally, the Air Force Special Operations Command conceded that it was their aircraft and that they were doing "training," but would not say more.[102]

It is unknown how often the military deploys mass aerial surveillance over American cities, but as Michel put it, "wide-area-camera manufacturers and users often turn the all-seeing eye on peacetime populations in the United States and elsewhere without their knowledge."[103] Often that surveillance is done with manned aircraft, as in Seattle. But in August 2019, *The Guardian* discovered that the Pentagon was testing up to 25 unmanned solar-powered balloons that were designed to "provide a persistent surveillance system to locate and deter narcotic trafficking and homeland security threats." The balloons, which traveled across six midwestern states at an altitude of 60,000 feet, were manufactured by the Sierra Nevada Corporation, developer of the original Gorgon Stare WAMI system. According to FAA documents, the balloons were equipped with synthetic aperture radar—and also with video, raising the possibility that Gorgon Stare or a successor was also installed on the balloons and recording large swaths of the U.S. Midwest.[104]

The military has always tested equipment and techniques intended for overseas battlefield use in the United States, and it has always loaned its resources to civilian authorities (in strictly limited ways) for help in emergencies.[105] But when that testing or that civilian assistance involves the mass surveillance of American populations, everything changes. It raises important questions: for example, as WAMI technology spreads and loosening restrictions on drone flights provide a cheap platform for that technology, what will the future of military drone flights over the United States be, and what checks and balances will be imposed on those flights?

The Pentagon did not disclose its domestic use of drones until forced to do so by a Freedom of Information Act (FOIA) request in 2016.[106] "Pentagon admits it has deployed military spy drones over the U.S.," as *USA Today* reported.[107] The Pentagon disclosed 11 total domestic drone missions between 2011 and 2017—but then carried out 11 more in 2018.[108] Four of the latter missions were to support firefighters, and several others were for the military's own purposes. But one was to support "counter-drug operations"—a "mission" that lasted five months—and another was to "support civil authorities at the southern border"—a mission that consisted of being "On Call Throughout 2018."[109] Pentagon officials also considered deploying additional military drones on the U.S.-Mexico border in 2019, according to documents obtained by *Newsweek*, as part of President Donald Trump's deployment of federal military forces to the border. The deployment was considered to fill a shortage in border equipment created after Democratic state governors pulled their state militias out of border duty amid controversy over the Trump administration's border practices.[110]

Some defense experts have recommended that the Air Force push toward a total shift to drones as an alternative to manned aircraft.[111] The U.S. military already possesses many thousands of drones[112] and is investing heavily in the technology, requesting $9.4 billion for UAS in the fiscal year 2019 budget.[113] And in recent years, the Pentagon has been pushing hard to overcome the FAA's barriers on its domestic large-drone flights.

Those barriers are significant. In the FAA Modernization and Reform Act of 2012, Congress called for the integration of drones into the national airspace by 2015.[114] But that timeline appears to have been wildly unrealistic. It remains true, as the GAO put it, that "no routine operations—meaning those that can occur without any prior authorization—are currently allowed for large UAS (55 pounds and over) for any purpose."[115] Anyone wanting to fly a large drone (or a drone above an altitude of 400 feet, or in violation of any of the FAA's other strictures on small drones) must seek permission (called a Certificate of Authorization, or COA) on a case-by-case basis. Unlike small drones, large drones usually need to fly beyond the visual line of sight of their operators (though in some testing, the FAA requires that UAVs be followed by a manned chase plane).

One reason for the FAA's caution is obviously that the consequences of an accident involving a large drone are potentially much more severe than an accident involving a small one. A 2019 report found more than 250 crashes around the world involving large drones in the previous decade.[116] In 2014, the *Washington Post* reported that since 2001 there had been at least 49 major military drone crashes in the United States—a number that did not include near-misses or any crashes that inflicted less than $2 million in damages. These accidents included the crash of a drone the size of a Boeing 767 in rural Maryland in 2012 and the crash of a 375-pound drone onto an elementary school playground in Pennsylvania. Reporters at the *Washington Post* had to file FOIA requests because the Pentagon and FAA refused to share information on the crashes with the public.[117]

It's important to understand that the military has a different attitude toward risk than civilians do. The military inherently understands that nothing is without risk, and it is accustomed to seeing things in terms of tradeoffs. As one official put it, "With the military, everything is a risk decision."[118] On a battlefield, the risk that an aircraft might crash has to be balanced against other risks, such as the risk that enemies will go undetected or that lifesaving operations won't take place. In the domestic aviation sphere, however, authorities find almost no risk acceptable and think in terms of eliminating it entirely (something they have very nearly succeeded in doing with large commercial airliners). The risk that a poorly controlled UAV will come crashing down on the heads of soldiers operating in a war zone in Afghanistan cannot be treated the same way as the risk that such a drone will crash into a schoolyard or a passenger airliner. As military flights come home, there is a risk that that difference in attitude will have tragic consequences.

Despite the barriers and obstacles, pro-drone forces in Congress have consistently pushed for the integration of drones, and military drones in particular, into the national airspace. As a result, the military has been heavily involved in the FAA's effort to get drones to operate "harmoniously, side-by-side with manned aircraft."[119] Indeed some have criticized the heavy military presence in the push to integrate drones into civilian airspace, even on the civilian side.[120] In 2008, Congress declared that "the pace of progress . . . has been insufficient and poses a threat to national security."[121] The following year Congress directed the Pentagon and Department of Transportation to "jointly develop a plan for providing expanded access to the national airspace for unmanned aircraft systems of the Department of Defense."[122] In response, the Pentagon in 2010 issued a sprawling, multiyear plan to integrate military drones into the national airspace.[123] A centerpiece of the plan was the creation of a cross-governmental UAS Executive Committee to spearhead "increased—and ultimately routine—access of Federal UAS" in the national airspace.[124] The committee's membership consisted of two FAA officials, one official from DHS, one from NASA—and two from the Pentagon, ensuring that the military would play a central role in the ongoing effort to integrate drones into American aviation.[125]

A central part of the government's approach has been the creation of test sites where drones can fly outside the usual rules to test integration strategies. The FAA has given permission for large drones to fly BVLOS at altitudes up to 10,000 feet at a test site around Grand Forks Air Force Base in North Dakota, and for large drones to fly up to 18,000 feet within an area of approximately 15,000 square miles in New Mexico.[126] In 2018, the Pentagon also loosened its own rules for domestic drone flights, dispensing with a requirement that the secretary of defense personally approve all domestic drone missions and allowing lower-ranking officials to approve such missions.[127]

Of course innumerable other U.S. military drone efforts are underway, aimed at expanding overseas military capabilities. But clearly, as with the CBP's use of the Reaper, such technologies have a way of finding their way into domestic use.

OTHER FEDERAL AGENCIES

In 2013, FBI Director Robert Mueller was answering questions about telephone data collection before the Senate Judiciary Committee when California Sen. Dianne Feinstein asked him about his agency's use of drones. Mueller replied that the agency used the technology in a "very, very minimal way." He said the technology was "very seldom used and . . . very narrowly focused on particularized cases and particularized needs."[128]

This was not surprising at the time, as the technology was just beginning to enter wide use, was even more tightly restricted by the FAA than it is today, and, given those restrictions, was not something that the FBI had an obvious need to use in broader ways.

Since then, however, it has become easier for federal agencies such as the FBI to use drones. No other federal agencies are using the technology to the same extent as the military and CBP, but there is little doubt that drone use is growing. The FAA has entered into memoranda of understanding with a number of other federal agencies to "enable each agency access to certain airspace for public aircraft operations in accordance with applicable laws and government agency policy." The FAA has signed these agreements with DHS, the Department of Justice, Department of the Interior, and U.S. Forest Service, among others.[129]

The FAA has expanded other avenues for government use of drones as well. In 2015, the FBI wanted to use a drone to investigate a shooting in Tennessee, but due to a miscommunication with the FAA was not able to do so. "We realized during that event that there was a gap in the way UAS were cleared for operation in the National Airspace," an FAA official recounted in 2019. As a result, the FAA created a standing, on-call group with authority to rapidly authorize emergency drone flights outside of the baseline rules.[130] After creation of that group (called the Systems Operations Support Center, or SOSC), the number of exemptions from normal drone flight rules granted to law enforcement and other officials nearly doubled.[131] These exceptions, called Special Government Interest authorizations, can be granted for large national events such as the Superbowl, emergencies such as hurricanes, and law enforcement uses including finding hit-and-run drivers.[132] Drones are also being used increasingly by the U.S. Forest Service and other agencies to fight fires.[133]

It's not clear whether any checks and balances are in place to prevent the misuse of the term "emergency," which has been a common government tactic throughout history. We do know that the FAA has been excessively deferential to law enforcement. In 2014, for example, the FAA approved a no-fly zone over Ferguson, Missouri, during protests there over the police shooting death of black teenager Michael Brown. The 37-square-mile restriction was put in place at the request of police, ostensibly to protect public safety; but documents and audio recordings indicate that the police really just wanted to keep news helicopters from recording their activities—and that FAA officials were aware of the real purpose. The FAA specifically tailored the restriction to keep out news helicopters while allowing other routine air traffic—in a context where law enforcement was actively working to suppress media coverage of police behavior during the protests.[134] These restrictions were a clear violation of the First Amendment.[135] The FAA also approved questionable flight restrictions over Dakota Access Pipeline protests in North Dakota in 2016 during significant and troubling law enforcement activity, such as the use of heavily militarized weapons against protesters during an October 2016 raid of the protesters' camp by hundreds of soldiers and police. In a spread out, rural setting where drones allow reporters to see what police are doing in a way that is not otherwise possible without expensive aircraft, the police shot down or confiscated the drones of several reporters.[136]

In short, it has become easier for government agencies to deploy drones since 2013. We also know that federal law enforcement agencies make regular use of aerial surveillance. In May 2015, several days after the death in police custody of a black man, Freddie Gray, and the protests that his death sparked, several Baltimore residents noticed that unidentified aircraft had been circling over the city for several nights in a row. It turned out that the manned aircraft were operated by the FBI, which said they were "used to assist in providing high-altitude observation of potential criminal activity to enable rapid response by police officers on the ground."[137] Given that every person alive could *potentially* engage in criminal activity, that statement hardly seemed reassuring, though the FBI denied it was engaged in mass surveillance. That October, FBI Director James Comey acknowledged that the bureau had also flown flights over Ferguson during the protests there.[138]

Several weeks after the Baltimore sighting, the Associated Press reported that the FBI was operating "a small air force with scores of low-flying planes across the U.S.," carrying video cameras and sometimes cell-site simulators. Over a 30-day period, the AP found, the bureau had flown above more than 30 cities across the United States, hiding the planes' identity by registering them to front companies.[139] In a FOIA request, the American Civil Liberties Union learned that the agency had used video cameras and specialized night-vision cameras, and had engaged in unspecified "other electronic surveillance" from the planes.[140] The FBI also released more than 18 hours of video taken over Baltimore during the protests there, showing that the bureau had captured hours of video of various protests, marches, and rallies in public places. The aerial cameras also sometimes followed individuals walking or driving through the city.[141]

An investigation by BuzzFeed shed more light on the routine nature and scope of the federal government's manned surveillance aircraft program: the government tracked more than 1,950 flights over a four-month observation period, including dozens of flights each weekday.[142]

All of this surveillance appears to have involved manned flights, but these revelations leave little doubt that the FBI would be very interested in making use of drones for aerial surveillance. In 2015, the Department of Justice (DOJ) inspector general (IG) released an audit of the department's use of drones. The FBI reported to the IG that between 2006 and 2014, it had deployed drones "to support 13 investigations by acquiring imagery" for various operations. The IG also found that the FBI had used drones supplied by CBP 13 additional times. All of the deployments, the IG found, were done with FAA permission and "exclusively to provide targeted aerial surveillance in the context of specific ongoing investigations."[143]

Relatively little information is available about the use of drones by other federal law enforcement agencies. The DOJ inspector general reported that not only the FBI but also three other agencies—the ATF, the Drug Enforcement Administration, and the U.S. Marshals Service—"received support from" CBP and its Predators. The IG found that the ATF spent around $600,000 on drones, only to find that the machines were technically inadequate, and never flew them.[144]

It is unknown how the FBI or other law enforcement agencies may be planning to expand their use of drones to carry out the aerial surveillance they already conduct on a daily basis across the United States. In November 2019, DOJ did issue an updated internal policy on the use of UAS.[145] The policy contains good-sounding language about the importance of privacy and civil liberties, but in addition to lacking the force of law, the policy is so broad that the extent to which it will actually constrain federal law enforcement remains unclear. For example, it says that in determining how to use aerial sensors, DOJ personnel "will assess the potential intrusiveness and impact on privacy and civil liberties, which will be balanced against the relevant governmental interests." The policy also says that DOJ "will only use UAS in connection with properly authorized investigations and activities."[146] None of that language can be counted on to preclude the FBI or another agency from carrying out extended wide-area surveillance.

Public, media, and legislative interest in government uses of drones peaked around 2015, as public anticipation over the integration of drones in American life far outpaced reality. Today, with visible drone integration continuing at a slow pace, media attention is much more focused on the potential for drones to be used by hostile actors; and the flow of public information about government use of drones being sought from and released by federal agencies has correspondingly shrunk. The need for transparency and public discussion over drone technology has never waned, however. As shown, the wheels are in motion for this technology—and its use for surveillance—to become far more powerful and far more prevalent than it is today.

ACKNOWLEDGMENTS

The book you are reading is the product of a group effort. Jason Kuznicki, the Cato Institute's book editor, first proposed the book and provided invaluable insight and suggestions throughout the editing process. Eleanor O'Connor, managing director of Cato Books, had the unenviable job of keeping the project on schedule and making sure that the book looked as well designed as it does. Special thanks to Karen Coda, Emma Evans, Marcy Gessel, Karen Ingebretsen, Kay McCarthy, and Lucy Williams of Publications Professionals LLC for the copyediting and proofreading. Not many editors are tasked with a manuscript that includes such a plethora of acronyms, federal regulation citations, and references to court decisions spanning from medieval England to the present day.

Thanks finally to the contributors. I appreciate their willingness to contribute to the volume, which I believe will inform many timely and important discussions in the age of the drone.

NOTES

INTRODUCTION

1. *Florida v. Riley,* 488 U.S. 445, 462 (1989) (Brennan, J., dissenting).

CHAPTER ONE

1. See, for example, "Uber Elevate Urban Air Mobility Summit 2019," Speeches, Federal Aviation Administration (FAA), updated June 11, 2019, https://www.faa.gov/news/speeches/news_story.cfm?newsId=23794.

2. Federal Aviation Administration, Advisory Circular 91–57, June 9, 1981, https://www.faa.gov/documentLibrary/media/Advisory_Circular/91-57.pdf.

3. See, for example, 14 C.F.R. §§ 25.851, 121.139 (2016).

4. 14 C.F.R. § 91.119(b), (c) (1989).

5. See, for example, 14 C.F.R. § 119.31 *et seq.*; 14 C.F.R. §§ 135.247, 135.293 (2016); 14 C.F.R. § 61.12 (2009).

6. See, for example, 14 C.F.R. § 91.903 (1989).

7. National Defense Authorization Act for Fiscal Year 2012, Pub. L. No. 112–81, § 1097(a), 125 Stat. 1298 (2011).

8. National Defense Authorization Act for Fiscal Year 2012, § 1097(b).

9. Federal Aviation Administration, "FAA Selects Unmanned Aircraft Systems Research and Test Sites," press release, December 30, 2013.

10. FAA Modernization and Reform Act of 2012, Pub. L. No. 112–95, § 332(a)(1), 126 Stat. 11 (2012) (FMRA).

11. FAA Modernization and Reform Act of 2012, § 332(a)(5).

12. FAA Modernization and Reform Act of 2012, § 332(b)(1).

13. FAA Modernization and Reform Act of 2012, § 332(b)(2). Section 332 also recodified the test range requirement from the 2012 NDAA, FAA Modernization and Reform Act of 2012, § 332(c); required the FAA to designate permanent areas in the Arctic where UAS may be operated for research and commercial purposes, FAA Modernization and Reform Act of 2012, § 332(d)(1); required the FAA to update its policy statement on UAS, FAA Modernization and Reform Act of 2012, § 332(b)(3); and required various reports to Congress, FAA Modernization and Reform Act of 2012, § 332(a)(4), (c)(5).

14. FAA Modernization and Reform Act of 2012, § 333.

15. FAA Modernization and Reform Act of 2012, § 333(b)(2).

16. Type certificates, production certificates, airworthiness certificates, and design and production organization certificates, 49 U.S.C. § 44704(d), (a) (2018).

17. 49 U.S.C. § 44711(a)(1).

18. 14 C.F.R. §§ 21, 43, 45, 47, 61, 91, 101, 107, 183, https://www.faa.gov/regulations_policies/rulemaking/recently_published/media/2120-aj60_nprm_2-15-2015_joint_signature.pdf at 55.

19. FAA Modernization and Reform Act of 2012, § 334(a)(2).

20. FAA Modernization and Reform Act of 2012, § 334(c)(1).

21. FAA Modernization and Reform Act of 2012, § 336(a)(1).

22. FAA Modernization and Reform Act of 2012, §§ 336(a)(2), (a)(3).

23. FAA Modernization and Reform Act of 2012, § 336(c).

24. FAA Modernization and Reform Act of 2012, § 336(b).

25. See, for example, Peter H. Curtis, "A Model Airplane Club Grounded in Bladensburg," *Washington Post,* December 7, 2012.

26. See, for example, State Farm exemption No. 11188, available at https://www.regulations.gov/document?D=FAA-2014-0856-0007.

27. See Section 333 Authorizations Granted, Federal Aviation Administration, last modified September 20, 2016, https://www.faa.gov/uas/advanced_operations/certification/section_44807/authorizations_granted/.

28. Anna M. Gomez, "FAA Streamlines Airspace Authorizations for Section 333 Exemption Holders," Wiley, March 24, 2015.

29. Anna M. Gomez, "FAA Moves to Summary Grant Process for UAS Petitions," Wiley, April 14, 2015.

30. Anna M. Gomez, "FAA Revises Section 333 Grant Terms to Give Commercial UAS Operators Greater Flexibility," Wiley, March 14, 2016.

31. FMRA § 332(b)(1).

32. The Administrative Procedure Act requires administrative agencies to publish notices of proposed rulemaking in the Federal Register prior to adopting new regulations. Under the statute, such notices must include "(1) a statement of

the time, place, and nature of public rule making proceedings; (2) reference to the legal authority under which the rule is proposed; and (3) either the terms or substance of the proposed rule or a description of the subjects and issues involved." 5 U.S.C. § 553(b) (1966).

33. 14 C.F.R. 107 §§ 107.51(b) (altitude limitation), 107.51(a) (speed limit), 107.51(c)–(d) (visibility requirements), 107.29 (prohibition on nighttime operation), 107.39 (prohibition on operations over people), 107.31 (prohibition on operations beyond visual line of sight) (2016).

34. Compare 14 C.F.R. §§ 61.121–61.133 with 14 C.F.R. §§ 107.53–107.79 (2016).

35. Federal Aviation Administration, "FAA Hits 100K Remote Pilot Certificates Issued," last modified July 26, 2018.

36. 49 U.S.C. § 40102(a)(5), (a)(25). Courts have applied the common law meaning of common carriage in the air carrier context, finding that to be a common carrier, an entity must indiscriminately "hold itself out to the public"—or, in some jurisdictions, to a definable segment thereof—as willing to provide air transportation for hire. See, for example, *CSI Aviation Services Inc. v. U.S. Department of Transportation,* 637 F.3d 408, 415 (D.C. Cir. 2011) ("'Common carrier' is a well-known term that comes to us from the common law. . . . The term refers to a commercial transportation enterprise that 'holds itself out to the public' and is willing to take all comers who are willing to pay the fare, 'without refusal.'") (citing *Black's Law Dictionary,* 8th ed. (2004), p. 226); *Woolsey v. National Transportation Safety Board,* 993 F.2d 516, 523 (5th Cir. 1993) ("[T]he crucial determination in assessing the status of a carrier is whether the carrier has held itself out to the public or to a definable *segment* of the public as being willing to transport for hire, indiscriminately.") (emphasis in original) (citations omitted).

37. Air carrier operating certificates, 49 U.S.C. § 44705 (1994).

38. 49 U.S.C. § 44704.

39. 49 U.S.C. §§ 41101, 41112, 41501.

40. 49 U.S.C. § 44701(d)(1); see, for example, 14 C.F.R. Ch. I, subchapters C, G.

41. 14 C.F.R. § 107.1 (2016).

42. Operation and Certification of Small Unmanned Aircraft Systems, Final Rule, 81 Fed. Reg. 42064, 42076, June 28, 2016, https://www.govinfo.gov/content/pkg/FR-2016-06-28/pdf/2016-15079.pdf ("Final Part 107 Rule"). Final rule at 49.

43. 81 Fed. Reg. at 42076.

44. *Central West Virginia Regional Airport Authority Inc. v. Triad Engineering Inc.,* No. 2:15-CV-11818, 2016 WL 685086, at *9 (S.D. W.Va. February 18, 2016) (citing *United States v. City of Montgomery,* 201 F. Supp. 590, 593 (M.D. Ala. 1962) (finding the term "air carrier" to include municipality that runs an airport and companies operating restaurants at an airport); *City of Philadelphia v. C.A.B.,* 289 F.2d 770, 774

(D.C. Cir. 1961) (holding that trucking portion of shipping route constitutes "air transportation" when another part of the route is completed by air).

45. *Federal Express Corporation v. California Public Utilities Commission,* 936 F.2d 1075, 1079 (9th Cir. 1991).

46. 936 F.2d at 1078–79 (quoting 49 U.S.C. App. § 1305(a)(1) (1988)).

47. Office of the Secretary of Transportation, "Operation and Certification of Small Unmanned Aircraft Systems," June 21, 2016, https://www.faa.gov/uas/media/RIN_2120-AJ60_Clean_Signed.pdf at 31.

48. Waiver policy and requirements, 14 C.F.R. § 107.200 (2016).

49. "Waiver Safety Explanation Guidelines for Part 107 Waiver Applications," Part 107 Waivers, Federal Aviation Administration, last modified December 21, 2018. https://www.faa.gov/uas/commercial_operators/part_107_waivers/waiver_safety_explanation_guidelines/.

50. "Airspace & Airspace Authorizations" Webinars, Federal Aviation Administration, last modified July 22, 2020, https://www.faa.gov/uas/resources/webinars/archive/.

51. See "Part 107 Waivers Issued," Federal Aviation Administration, last modified June 10, 2021, https://www.faa.gov/uas/commercial_operators/part_107_waivers/waivers_issued/.

52. See, for example, Federal Aviation Administration, "Certificate of Waiver Authorization," https://www.faa.gov/uas/commercial_operators/part_107_waivers/waivers_issued/media/107W-2017-00003_Cory_Wise_CoW.pdf.

53. U.S. Department of Transportation, Office of Inspector General, "Opportunities Exist for FAA to Strengthen Its Review and Oversight Processes for Unmanned Aircraft System Waivers," Report No. AV2019005 at 8, November 7, 2018, https://www.oig.dot.gov/sites/default/files/FAA%20UAS%20Waivers%20Final%20Report%5E11-07-18.pdf at 8.

54. U.S. Department of Transportation, Report No. AV2019005 at 9.

55. U.S. Department of Transportation, Report No. AV2019005 at 10.

56. U.S. Department of Transportation, Report No. AV2019005 at 24.

57. U.S. Department of Transportation, Report No. AV2019005 at 24.

58. Final Part 107 Rule, 81 Fed. Reg. at 42071.

59. 81 Fed. Reg. at 42071.

60. Operation and Certification of Small Unmanned Aircraft Systems, Notice of Proposed Rulemaking, 80 Fed. Reg. 9544, 9556–9558, February 23, 2015, https://www.govinfo.gov/content/pkg/FR-2015-02-23/pdf/2015-03544.pdf.

61. 80 Fed. Reg. at 9557–58. Although the agency would not require a knowledge test, it would require the applicant to self-certify familiarity with relevant areas of knowledge. 80 Fed. Reg. at 9557.

62. Final Part 107 Rule, 81 Fed. Reg. at 42123.

63. 81 Fed. Reg. at 42123.

64. Micro Unmanned Aircraft Systems Aviation Rulemaking Committee (ARC), ARC Recommendations: Final Report, at 4, April 1, 2016, https://www.faa.gov/uas/resources/policy_library/media/Micro-UAS-ARC-FINAL-Report.pdf.

65. ARC, April 1, 2016 Final Report, at 6–7.

66. ARC, April 1, 2016 Final Report, at 9. Category 2 UAS would include those above 250 grams presenting a 1 percent or less chance of serious injury in the event of impact and would be required to meet minimum stand-off distances. ARC, April 1, 2016 Final Report, at 7–8. Category 3 UAS would include those above 250 grams presenting a 30 percent or less chance of serious injury in the event of impact but do not require sustained flights over crowds, and would be required to meet minimum stand-off distances and operate over people only if the site is closed or access is restricted, or operation over people is transient. ARC, April 1, 2016 Final Report, at 8, 12.

67. Exec. Order No. 12,866, 58 Fed. Reg. 51735 (1993), https://www.archives.gov/files/federal-register/executive-orders/pdf/12866.pdf.

68. "Drones: A Story of Revolution and Evolution," Speeches, Federal Aviation Administration, last modified January 6, 2020, https://www.faa.gov/news/speeches/news_story.cfm?newsId=21316.

69. FAA Extension, Safety, and Security Act of 2016, Pub. L. No. 114-90, § 2202, 130 Stat. 615, https://www.congress.gov/114/plaws/publ190/PLAW-114publ190.pdf ("2016 Extension Act").

70. ARC Recommendations: Final Report, UAS Identification and Tracking (UAS ID) Aviation Rulemaking Committee (ARC), at 1, September 30, 2017, https://www.faa.gov/regulations_policies/rulemaking/committees/documents/media/UAS%20ID%20ARC%20Final%20Report%20with%20Appendices.pdf.

71. ARC, September 30, 2017 Final Report, at 1–2.

72. ARC, September 30, 2017 Final Report, at 3.

73. ARC, September 30, 2017 Final Report, at 5. The ARC made a number of other recommendations covering issues such as operations near airports and critical infrastructure, stages of implementation of remote ID and tracking rules, and the role of air traffic control in remote ID systems. See ARC, September 30, 2017 Final Report, at 31–46.

74. ARC, September 30, 2017 Final Report, at 31, Appendix B at 17, 27.

75. ARC, September 30, 2017 Final Report, at Appendix D, Responses and Voting Results, at 27.

76. FAA Modernization and Reform Act of 2012, § 336(a) (emphasis added).

77. Registration and Marking Requirements for Small Unmanned Aircraft, 80 Fed. Reg. 78594 (December 16, 2015), https://www.govinfo.gov/content/pkg/FR-2015-12-16/pdf/2015-31750.pdf.

78. 5 U.S.C. 553(b)(B) (1966).

79. Registration and Marking Requirements for Small Unmanned Aircraft, Interim Final Rule, 80 Fed. Reg. at 78643, 78597 (December 16, 2015), https://www.govinfo.gov/content/pkg/FR-2015-12-16/pdf/2015-31750.pdf.

80. 80 Fed. Reg. at 78599.

81. *Taylor v. Huerta*, 856 F.3d 1089 (D.C. Cir. 2017). Taylor also challenged an advisory circular that prohibited the operation of model aircraft in certain restricted areas; that challenge was found to be untimely. *Taylor*, 856 F.3d at 1091.

82. *Taylor*, 856 F.3d at 1093.

83. *Taylor*, 856 F.3d at 1093. Because the text squarely barred the FAA from imposing the registration requirements on hobbyists, the court also found unavailing the FAA's argument that the rule was permissible because it was "consistent with one of the general directives of the [FMRA]: to 'improve aviation safety.'"

84. National Defense Authorization Act for Fiscal Year 2018. Pub. L. No. 115–9113191, 131 Stat. 1283, https://www.congress.gov/115/bills/hr2810/BILLS-115hr2810enr.pdf.

85. U.S. Department of Transportation, "Presidential Memorandum for the Secretary of Transportation," Sec. 2, last updated, October 31, 2017, https://www.transportation.gov/briefing-room/presidential-memorandum-secretary-transportation.

86. Federal Aviation Administration, "Completed Programs and Partnerships," last modified, October 27, 2018, https://www.faa.gov/uas/programs_partnerships/completed/.

87. See Federal Aviation Administration, *Certificate of Waiver or Authorization,* 107W-2016-00001A, Atlanta, Georgia, 2016, https://www.faa.gov/uas/commercial_operators/part_107_waivers/waivers_issued/media/107W-2016-00001A_CNN_CoW.pdf (waiver authorizing CNN to operate over people); Federal Aviation Administration, *Certificate of Waiver or Authorization,* 107W-2016-00003, Fort Worth, Texas, 2016, https://www.faa.gov/uas/commercial_operators/part_107_waivers/waivers_issued/media/107W-2016-00003_BNSF_CoW.pdf (waiver authorizing BNSF to operate BVLOS); Federal Aviation Administration, *Certificate of Waiver or Authorization,* 107W-2016-00002, Raleigh, North Carolina, 2016, https://www.faa.gov/uas/commercial_operators/part_107_waivers/waivers_issued/media/107W-2016-00002_PrecisionHawk_CoW.pdf (waiver authorizing Precision Hawk to operate BVLOS).

88. AUVSI letter to the President-UAS Pilot Program, October 11, 2017, AUVSI.org, https://www.auvsi.org/sites/default/files/PDFs/AUVSI_Letter-to-the-President_UAS-Pilot-Program-FINAL.PDF.

89. Unmanned Aircraft Systems Integration Pilot Program—Announcement of Establishment of Program and Request for Applications, 82 Fed. Reg. 51903,

51904, November 8, 2017, https://www.govinfo.gov/content/pkg/FR-2017-11-08/pdf/2017-24126.pdf.

90. 82 Fed. Reg. at 51905.

91. U.S. Department of Transportation, "U.S. Transportation Secretary Elaine L. Chao Announces Unmanned Aircraft Systems Integration Pilot Program Selectees," press release, May 9, 2018.

92. Federal Aviation Administration, "BEYOND," https://www.faa.gov/uas/programs_partnerships/beyond/.

93. Request for Emergency Processing of Collection of Information by the Office of Management and Budget; Emergency Clearance to Revise Information Collection 2120–0768, Part 107 Authorizations and Waivers, 82 Fed. Reg. 47289, 47290, October 11, 2017, https://www.govinfo.gov/content/pkg/FR-2017-10-11/pdf/2017-21878.pdf ("Emergency LAANC Request").

94. P. Jay Merkle et al., "Low Altitude Authorization and Notification Capability (LAANC) Concept of Operations," Federal Aviation Administration, May 12, 2017, https://www.faa.gov/uas/programs_partnerships/data_exchange/laanc_for_industry/media/laanc_concept_of_operations.pdf ("LAANC CONOPS").

95. See Merkle et al., "LAANC CONOPS," at 19.

96. Emergency LAANC Request, 82 Fed. Reg. at 47290.

97. Public information collection activities; submission to Director; approval and delegation, 44 U.S.C. § 3507. (2000).

98. Emergency processing, 5 C.F.R. § 1320.13(a)(2).

99. Emergency LAANC Request, 82 Fed. Reg. at 47290.

100. Office of Information and Regulatory Affairs, ATC Authorization in Controlled Airspace under Part 107 (Reginfo.gov, 2017), https://www.reginfo.gov/public/do/PRAViewICR?ref_nbr=201710-2120-001.

101. "FAA Fields Prototype UAS Airspace Authorization System," Federal Aviation Administration, last modified November 16, 2017, https://www.faa.gov/news/updates/?newsId=89186.

102. "FAA Begins Drone Airspace Authorization Expansion," Federal Aviation Administration, last modified April 30, 2018, https://www.faa.gov/news/updates/?newsId=90245.

103. "More than 50,000 LAANC Applications Processed," Federal Aviation Administration, last modified November 19, 2018, https://www.faa.gov/news/updates/?newsId=92273&omniRss=news_updatesAoc&cid=101_N_U.

104. "LAANC Drone Program Expansion Continues," Federal Aviation Administration, last modified December 2, 2019, https://www.faa.gov/news/updates/?newsId=94750.

105. See, for example, Conor Dougherty, "Drone Developers Consider Obstacles That Cannot Be Flown Around," *New York Times,* September 1, 2014; Trevor

Mogg, "Amazon Drone Delivery Plan Given Hope as NASA Progresses with Air Traffic Control System," Digital Trends Media Group, March 13, 2015; Bill Carey, "NASA, FAA Discuss Formal Collaboration on Small Drones," *AIN Online*, September 17, 2015.

106. Merkle et al., "LAANC CONOPS" at 7.

107. FAA Reauthorization Act of 2018, Pub. L. No. 115–254, 132 Stat. 3186, (2018), https://www.congress.gov/115/plaws/publ254/PLAW-115publ254.pdf ("2018 FAA Reauthorization Act").

108. 2018 FAA Reauthorization Act, § 349(a). The full list of operating parameters in Section 349 is as follows: "(1) The aircraft is flown strictly for recreational purposes; (2) The aircraft is operated in accordance with or within the programming of a community-based organization's set of safety guidelines that are developed in coordination with the Federal Aviation Administration; (3) The aircraft is flown within the visual line of sight of the person operating the aircraft or a visual observer co-located and in direct communication with the operator; (4) The aircraft is operated in a manner that does not interfere with and gives way to any manned aircraft; (5) In Class B, Class C, or Class D airspace or within the lateral boundaries of the surface area of Class E airspace designated for an airport, the operator obtains prior authorization from the Administrator or designee before operating and complies with all airspace restrictions and prohibitions; (6) In Class G airspace, the aircraft is flown from the surface to not more than 400 feet above ground level and complies with all airspace restrictions and prohibitions; (7) The operator has passed an aeronautical knowledge and safety test described in subsection (g) and maintains proof of test passage to be made available to the Administrator or law enforcement upon request; (8) The aircraft is registered and marked in accordance with chapter 441 of this title and proof of registration is made available to the Administrator or a designee of the Administrator or law enforcement upon request."

109. 2018 FAA Reauthorization Act, § 349(f). In addition to repealing and replacing Section 336, the Reauthorization Act required the FAA to develop the aeronautical knowledge test to be taken by recreational operators and publish an advisory circular on the process for recognizing community-based organizations, and authorized the FAA to establish rules, procedures, and standards to facilitate UAS use by institutions of higher education for educational or research purposes. See 2018 FAA Reauthorization Act, §§ 349, 350.

110. See 2018 FAA Reauthorization Act, §§ 376, 377.

111. 2018 FAA Reauthorization Act, § 352.

112. 2018 FAA Reauthorization Act, §§ 366, 371, 372, 379.

113. 2018 FAA Reauthorization Act, § 345.

114. 2018 FAA Reauthorization Act, § 369.

115. 2018 FAA Reauthorization Act, §§ 343, 344.

116. 2018 FAA Reauthorization Act, § 348.

117. 2018 FAA Reauthorization Act, §§ 358, 379.

118. 2018 FAA Reauthorization Act, §§ 346, 353, 359, 368, 379.

119. 2018 FAA Reauthorization Act, § 373.

120. 2018 FAA Reauthorization Act, § 374.

121. 2018 FAA Reauthorization Act, § 360.

122. 2018 FAA Reauthorization Act, § 342.

123. 2018 FAA Reauthorization Act, § 1602(b).

124. National Defense Authorization Act for Fiscal Year 2017, Pub. L. No. 114-328, 130 Stat. 2756, §§ 1697, 3112 (2016); National Defense Authorization Act for Fiscal Year 2018, Pub. L. No. 115-91, 131 Stat. 1283, § 1692 (2017).

125. Sara Baxenberg and Joshua Turner, "Is the US Ready for a 'Gatwick Drone' Scenario?," *Law 360*, January 8, 2019.

126. 2018 FAA Reauthorization Act, § 362(1).

127. 2018 FAA Reauthorization Act, § 383.

128. 2018 FAA Reauthorization Act, § 384. The act also introduced new civil and criminal penalties for other UAS-related conduct, including criminalizing knowing and willful operation of UAS within or above a "restricted building or grounds," 2018 FAA Reauthorization Act, § 381, criminalizing knowing or reckless interference with wildfire suppression or related law enforcement or emergency response activities, 2018 FAA Reauthorization Act, § 382, and imposing a civil penalty for equipping a UAS with a dangerous weapon, 2018 FAA Reauthorization Act, § 363.

129. See 14 C.F.R. Part 135 (1975). Part 135 establishes a five-phase process for air carrier certification. See also FAA, Completing the Certification Process, https://www.faa.gov/licenses_certificates/airline_certification/air_carrier/complete_cert_process/.

130. "FAA Begins Drone Airspace Authorization Expansion," Federal Aviation Administration, last modified April 30, 2018, https://www.faa.gov/news/updates/?newsId=90245.

131. Federal Aviation Administration, "U.S. Transportation Secretary Elaine L. Chao Announces FAA Certification of Commercial Package Delivery," press release, April 23, 2019.

132. See 2018 FAA Reauthorization Act, § 347.

133. U.S. Department of Transportation, Federal Aviation Administration, Exemption No. 18163, Regulatory Docket No. FAA-2018-0835 2018(2019), https://www.regulations.gov/contentStreamer?documentId=FAA-2018-0835

-0020&attachmentNumber=1&contentType=pdf, at 3 ("Wing Exemption"); U.S. Department of Transportation, Federal Aviation Administration, Exemption No. 18339, Regulatory Docket No. FAA-2019-0628 (2019), https://www.regulations.gov/contentStreamer?documentId=FAA-2019-0628-0020&attachmentNumber=1&contentType=pdf,at 2 ("UPS Exemption").

134. Wing Exemption at 6; UPS Exemption at 5.

135. *AUVSI News,* "Wing Completes Landmark UAS Package Delivery in Montgomery County, Virginia," *AUVSI*, August 8, 2018, https://www.auvsi.org/industry-news/wing-completes-landmark-uas-package-delivery-montgomery-county-virginia.

136. "UPS Partners with Matternet to Transport Medical Samples via Drone Across Hospital System in Raleigh, N.C.," UPS Pressroom, March 26, 2019, https://pressroom.ups.com/pressroom/ContentDetailsViewer.page?ConceptType=PressReleases&id=1553546776652-986.

137. Wing Exemption at 70; UPS Exemption at 33.

138. Notification to UAS Operators Proposing to Engage in Air Transportation, 83 Fed. Reg. 18734, 18735 (April 30, 2018), https://www.govinfo.gov/content/pkg/FR-2018-04-30/pdf/2018-09057.pdf.

139. 49 U.S.C. § 40103(a) (1994).

140. See 49 U.S.C. § 40103(b)(2) (directing the FAA to impose regulations related to "navigating, protecting, and identifying aircraft," "protecting individuals and property on the ground," "using the navigable airspace efficiently," and "preventing collision between aircraft . . . and airborne objects").

141. Federal Aviation Administration Office of the Chief Counsel, "State and Local Regulation of Unmanned Aircraft Systems (UAS) Fact Sheet," December 17, 2015, https://www.faa.gov/uas/resources/policy_library/media/UAS_Fact_Sheet_Final.pdf .

142. FAA, December 17, 2015 Fact Sheet at 2 (citing *Montalvo v. Spirit Airlines,* 508 F.3d 464 (9th Cir. 2007), and *French v. Pan Am Express, Inc.* 869 F.2d 1 (1st Cir. 1989); see also *Arizona v. U.S.* 567 U.S. 387, 401, 132 S. Ct. 2492, 2502 (2012)).

143. FAA, December 17, 2015 Fact Sheet at 3.

144. FAA, December 17, 2015 Fact Sheet at 3.

145. *Singer v. City of Newton*, 284 F. Supp. 3d 125, 128 (D. Mass. *appeal dismissed,* 2017), No. 17-2045, 2017 WL 8942575 (1st Cir. December 7, 2017).

146. 284 F. Supp. 3d at 128.

147. 284 F. Supp. 3d at 130–33.

148. *National Press Photographers Association v. McCraw*, No. 1:19-CV-946-RP, 2020 WL 7029159, at *13–15 (W.D. Tex. November 30, 2020).

149. 2020 WL 7029159, at *20, *22.

150. National Conference of Commissioners on Uniform State Laws, Tort Law Relating to Drones Act, June 19, 2018, https://www.uniformlaws.org/HigherLogic/System/DownloadDocumentFile.ashx?DocumentFileKey=acc3dee9-7ab6-24e3-ee55-ad3e09a82c6a&forceDialog=0.

151. See Restatement (Second) of Torts § 159(2). The committee's initial proposal appears to stem from a misunderstanding of *United States v. Causby*, a 1946 Supreme Court case in which the Court held that overflights by large military aircraft constituted a taking of a property owner's chicken farm. Although the court discussed the need for property owners to "control" the "immediate reaches" of their property, the court's holding depended on its determination that the overflights at issue were "so low and so frequent as to be a direct and immediate interference with the enjoyment and use of the land." *United States v. Causby*, 328 U.S. 256, 266, 66 S. Ct. 1062, 1068, 90 L. Ed. 1206 (1946). For more discussion about how the ULC's proposal is inconsistent with governing case law, see S. Baxenberg and J. Turner, "*Causby* and Effect: How the Uniform Law Commission's Misplaced Reliance on a 1946 Supreme Court Case Threatens the Drone Industry," *AUSVI's Unmanned Systems*, September 26, 2018.

152. U.S. Department of Transportation, *RE: Tort Law Relating to Drones Act–Comments Section*, letter, p. 2, July 11, 2018, https://www.uniformlaws.org/HigherLogic/System/DownloadDocumentFile.ashx?DocumentFileKey=21ef3370-77ea-5d39-cdb9-8d7816698cac&forceDialog=0.

153. National Conference of Commissioners on Uniform State Laws, "Uniform Tort Law Relating to Drones Act, Draft for Approval," Anchorage, Alaska, July 12–18, 2019, https://www.uniformlaws.org/HigherLogic/System/DownloadDocumentFile.ashx?DocumentFileKey=a870dac3-db42-85cd-197d-abbf2d32e30d&forceDialog=0.

154. See letter from the Joint Editorial Board for Uniform Real Property Acts to Commissioners, Uniform Law Conference (June 5, 2019), https://www.uniformlaws.org/HigherLogic/System/DownloadDocumentFile.ashx?DocumentFileKey=a22024a3-ad6c-5f2d-a96c-1e6981b32cba&forceDialog=0; letter from Jo-Ann Marzullo, Section Chair-Elect, Real Property, Trust & Estate Law Section, American Bar Association, to Commissioners, Uniform Law Conference (June 27, 2019), https://www.uniformlaws.org/HigherLogic/System/DownloadDocumentFile.ashx?DocumentFileKey=1f86816baff6-2b3a-5a3c-baf2f1a9146a&forceDialog=0; letter from American College of Real Estate Lawyers to Commissioners, Uniform Law Commission (July 3, 2019), https://www.uniformlaws.org/HigherLogic/System/DownloadDocumentFile.ashx?DocumentFileKey=170e1617-b926-16f6-1bb8-e6a222c375cf&forceDialog=0; see also letter from Henry E. Smith, Fessenden Professor of Law and Reporter, American Law Institute's Restatement Fourth of the Law, Property, to National Conference of Commissioners on Uniform State Laws (June 20, 2019), https://www.uniformlaws.org/HigherLogic/System

/DownloadDocumentFile.ashx?DocumentFileKey=44cb6696-a733-81f8-c9da-fb5c47852d59&forceDialog=0.

155. See Government Accountability Office, "Unmanned Aircraft Systems: FAA's Compliance and Enforcement Approach for Drones Could Benefit from Improved Communication and Data," End Note 20, October 17, 2019, https://www.gao.gov/reports/GAO-20-29 ("FAA officials told us DOT is currently reviewing the department's position regarding which types of state and local laws relating to UAS it believes may be federally preempted. The officials said they expect this review to be completed by the end of 2019, with results to be publicly announced as revised agency guidance or in some other form.").

156. Remote Identification of Unmanned Aircraft Systems, Notice of Proposed Rulemaking, 84 Fed. Reg. 72438, 72519–20 (December 31, 2019) (proposed § 89.310), https://www.govinfo.gov/content/pkg/FR-2019-12-31/pdf/2019-28100.pdf.

157. 84 Fed. Reg. at 72517–18 (proposed § 89.115), 72520-21 (proposed § 89.320).

158. 84 Fed. Reg. at 72519–20 (proposed §§ 89.305, 89.315).

159. Remote Identification of Unmanned Aircraft, Final Rule, 86 Fed. Reg. 4390, 4393, 4406 (January 15, 2021).

160. 86 Fed. Reg. at 4505–06 (Final Rules, 14 C.F.R. §§ 89.105, 89.110, 89.115).

161. 86 Fed. Reg. at 4506 (Final Rules, 14 C.F.R. § 89.115).

162. See 86 Fed. Reg. at 4509–10 (Final Rules, 14 C.F.R. §§ 89.515, 89.520).

163. See 86 Fed. Reg. at 4390.

164. Remote Identification of Unmanned Aircraft; Delay, Final rule; delay of effective and compliance date; correction, 86 Fed. Reg. 45 (March 10, 2021).

165. See 86 Fed. Reg. at 4390.

166. Operations of Small Unmanned Aircraft Systems over People, Final Rule, 86 Fed. Reg. 4314 (rel. January 15, 2021), https://www.govinfo.gov/content/pkg/FR-2021-01-15/pdf/2020-28947.pdf ("Flights over People and at Night Rules").

167. 86 Fed. Reg. at 4383 (Final Rules, 14 C.F.R. § 107.110).

168. 86 Fed. Reg. at 4383 (Final Rules, 14 C.F.R. § 107.120(a)).

169. 86 Fed. Reg. at 4383 (Final Rules, 14 C.F.R. §§ 107.110(c), 107.115(b)).

170. 86 Fed. Reg. at 4384 (Final Rules, 14 C.F.R. § 107.130).

171. 86 Fed. Reg. at 4384 (Final Rules, 14 C.F.R. § 107.125).

172. 86 Fed. Reg. at 4385 (Final Rules, 14 C.F.R. § 107.155).

173. 86 Fed. Reg. at 4384 (Final Rules, 14 C.F.R. § 107.140).

174. See, for example, Comments of the Association of Unmanned Vehicle Systems International, FAA Docket ID FAA-2018-1087-0487, at 10-13 (filed April 14, 2019), https://www.regulations.gov/document?D=FAA-2018-1087-0487 ("AUVSI Comments"); Comments of Kittyhawk.io, FAA Docket ID FAA-2018-1087-0821, at 1–2 (posted April 16, 2019), https://www.regulations.gov/document?D=FAA-2018-1087-0821 ("Kittyhawk Comments"); Comments of ParaZero, FAA Docket ID FAA-

2018-1087-0553, at 2 (posted April 15, 2019) https://www.regulations.gov/document?D=FAA-2018-1087-0553 ("ParaZero Comments"); Comments of PrecisionHawk Inc., FAA Docket ID FAA-2018-1087-0549, at 2-4 (filed April 14, 2019) https://www.regulations.gov/document?D=FAA-2018-1087-0549 ("PrecisionHawk Comments").

175. AUVSI Comments at 13–15; Kittyhawk Comments at 3; ParaZero Comments at 2; PrecisionHawk Comments at 3.

176. Flights over People and at Night Rules, 86 Fed. Reg. at 4385 (Final Rules, 14 C.F.R. § 107.145).

177. 86 Fed. Reg. at 4385.

178. 86 Fed. Reg. at 4382 (Final Rules, 14 C.F.R. § 107.29).

179. FAA, Executive Summary: Final Rule on Operation of Small Unmanned Aircraft Systems over People (December 28, 2020), p. 4.

180. Safe and Secure Operations of Small Unmanned Aircraft Systems, 84 Fed. Reg. 3732 (February 13, 2019).

181. 84 Fed. Reg. at 3734.

182. 84 Fed. Reg. at 3736.

183. 84 Fed. Reg. at 3736–37.

184. 84 Fed. Reg. at 3734–35.

185. 84 Fed. Reg. at 3735–36.

186. FAA Reauthorization Act of 2018, § 348.

187. Kelsey Reichmann, "First FAA Type Certification for Advanced Air Mobility Aircraft Could Come This Year," Aviation Today, January 26, 2021.

188. FAA Extension, Safety, and Security Act of 2016, § 2209.

189. FAA Extension, Safety, and Security Act of 2016, § 2209(b)(1)(C); 2018 FAA Reauthorization Act, § 369.

190. Office of Information and Regulatory Affairs, "Prohibit or Restrict the Operation of an Unmanned Aircraft in Close Proximity to a Fixed Site Facility," Fall 2020, https://www.reginfo.gov/public/do/eAgendaViewRule?pubId=202010&RIN=2120-AL33.

191. See, for example, Florida Unmanned Aircraft Systems Act, Fl. Stat. § 330.41 ("Florida UAS Law"); Texas Government Code, Title 4, Subtitle B, Ch. 423.

192. Florida UAS Law, § 330.419441(4)(e).

193. See *National Press Photographers Association et al. v. Steven McCraw et al.*, Civil Action No. 1:19-cv-00946 (W.D. Tex. 2019).

194. FAA, "NextGen Integration & Evaluation Capability: Projects," August 4, 2020.

195. 2016 Extension Act, § 2208.

196. 2018 FAA Reauthorization Act, § 376(a).

197. 2018 FAA Reauthorization Act, § 376(b).

198. 2018 FAA Reauthorization Act, § 376(c).

199. 2018 FAA Reauthorization Act, § 376(f).

200. 2018 FAA Reauthorization Act, § 377.

201. 2018 FAA Reauthorization Act, § 377(d).

202. See FCC, 2019 TAC Working Groups and Charter.

203. Commerce Spectrum Management Advisory Committee, NTIA, last modified July 29, 2020, https://www.ntia.doc.gov/category/csmac.

204. 2018 FAA Reauthorization Act, § 374.

205. Wireless Telecommunications Bureau, Office of Engineering and Technology, "Report on Section 374 of the FAA Reauthorization Act of 2018," August 20, 2020, p. 7. At present, a 2018 Petition for Rulemaking filed by the Aerospace Industries Association asking the FCC to establish technical and service rules to permit UAS command and control operations in the C-band remains pending before the agency. See Petition to Adopt Service Rules for Unmanned Aircraft Systems ("UAS") Command and Control in the 5030–5091 MHz Band, RM-11798 (filed February 8, 2018), https://ecfsapi.fcc.gov/file/1020998018431/AIA%20Petition%20for%20Rulemaking%20on%20UAS%202018-02-08%20FILED.pdf.

206. Wireless Telecommunications Bureau, "Report on Section 374," pp. 10–11.

CHAPTER TWO

1. What is a drone? As discussed more fully in the next section, it is simply an aircraft without a pilot on board. Regarding the use of the word "drone," let us consider a bit of nomenclature: unmanned aircraft or unmanned aerial vehicle refers to the aircraft itself. Unmanned aircraft system (UAS) refers to the aircraft plus the ground station and any external communication link. Another term, remotely piloted aircraft system, is too limited, because it does not include autonomous operations. The moniker "drone," first regarded as somewhat pejorative given the plain meaning of the word and its use as a lethal weapon in military operations, was rather quickly embraced by the UAS industry and the Federal Aviation Administration (FAA). For example, the FAA has a Drone Advisory Committee. Thus, "drone" is here to stay and will be used throughout this chapter.

2. 14 C.F.R. Part 107 (2016).

3. FAA, "UAS by the Numbers," as of April 19, 2021, https://www.faa.gov/uas/resources/by_the_numbers/.

4. FAA Advisory Circular 91-57 (June 9, 1981, canceled September 2, 2015). An Advisory Circular is not a rule; it only provides guidance to the public.

5. FAA "Unmanned Aircraft Systems Operations in the U.S. National Airspace System—Interim Operational Approval Guidance," Memorandum AFS-400, UAS Policy 05-01 (September 16, 2005).

6. "Unmanned Aircraft Operations in the National Airspace System," 72 Fed. Reg. 6689 (February 13, 2007).

7. "Simply stated, an unmanned aircraft is a device that is used, or intended to be used, for flight in the air with no onboard pilot." 72 Fed. Reg. at 6689.

8. *Huerta v. Pirker*, NTSB Order No. EA-5730, 2014 WL8095629.

9. Pub. L. No. 112-95 (February 14, 2012).

10. When the FAA issued an interpretive rule to implement Section 336, it received more than 40,000 comments. As of this writing, the FAA has not responded to those comments.

11. "[A] person may operate an aircraft only when the aircraft is registered under section 44103 of this title."

12. *Morton v. Mancari*, 417 U.S. 535, 549 (1974).

13. *Taylor*, 856 F. 3d at 1091.

14. Subsection 1092(d) of the National Defense Authorization Act of 2018, Pub. L. No. 115-91 (December 12, 2017).

15. Wayne Rosenkrans, "Fire Traffic Control," *AeroSafety World*, August 15, 2016.

16. Pub. L. No. 114-190, Section 2205.

17. Pub. L. No. 114-290, Section 2207. In the 2018 FAA Reauthorization Act, Congress reiterated its finding that drone flights by public and civil operators can help facilitate firefighting and other first responder activities and called upon the FAA to conduct a study of fire department and first responder use of drones. Pub. L. No. 115-254, Sections 353 and 359.

18. "Operation and Certification of Small Unmanned Aircraft Systems," 81 Fed. Reg. 42064 (June 28, 2016).

19. See Drone Advisory Committee, DAC Meeting, October 17, 2019, Washington: task group recommendation on Part 107 waivers.

20. Letter from Small UAV Coalition Aviation Counsel Gregory S. Walden to Secretary of Transportation Elaine Chao and FAA Administrator Steve Dickson, March 4, 2020, at https://smalluavcoalition.org/wp-content/uploads/2020/03/Small-UAV-Coalition-letter-to-DOT-Secretary-and-FAA-Admnistrator-030420.pdf. The coalition sent a follow-up letter on May 19, 2020, https://smalluavcoalition.org/wp-content/uploads/2020/05/small-UAV-Coalition-letter-to-DOT-and-FAA-on-COVID-19-response-05.19.20.pdf.

21. In relevant part, Section 91.137 authorizes the issuance of a temporary flight restriction to "provide a safe environment for the operation of disaster relief aircraft." Even if the FAA does not consider pandemic relief to be disaster relief, the FAA administrator has authority under 49 U.S.C. § 44701(f) to grant an exemption from a regulation such as Section 91.137 if "in the public interest." It is this statutory exemption authority that the Small UAV Coalition urged be used to authorize BVLOS delivery flights for compensation or hire (exempting operators

from 14 C.F.R. 107.205(c)). To date, the FAA has not elected to use its exemption authority in either instance.

22. Pub. L. No. 115-254 (October 5, 2018); 49 U.S.C. § 44807.

23. 14 C.F.R. 107.205(c).

24. 14 C.F.R. 135, Air Carrier and Operator Certification. FAA regulations contain several categories of air carrier. Part 121 is for scheduled airlines. Part 135 is mainly for on-demand, air taxi, or charter operations but also includes scheduled operations of aircraft with no more than nine seats.

25. At least three other companies are in the queue—Uber Elevate, Zipline International, and Causey Aviation—as the FAA has published the Section 44807 petitions and sought public comment.

26. "Operations of Small Unmanned Aircraft Systems over People," Final Rule, 86 Fed. Reg. 4314 (January 15, 2021).

27. Pub. L. No. 114-190 (July 15, 2016), Section 2202.

28. When the FAA published two rulemaking documents in February 2019, it explicitly stated that the OOP rulemaking proceeding would not reach a final rule until the remote ID final rule was in place. See "Operation of Small Unmanned Aircraft Systems over People," Notice of Proposed Rulemaking, FAA-2018-1087, 84 Fed. Reg. 3856, 3861 (February 13, 2019); "Safe and Secure Operations of Small Unmanned Aircraft Systems," Advance Notice of Proposed Rulemaking, FAA-2018-1086, 84 Fed. Reg. 3732, 3733–34 (February 13, 2019). In fact, federal government agencies had previously informed the UAS industry that no FAA UAS rule would proceed until two additional departments—Justice and Homeland Security—received statutory authority to deploy counter-UAS measures. That authority was granted in Section 1602 of the Preventing Emerging Threats Act of 2018, Pub. L. No. 115-254 (October 5, 2018).

29. "Remote Identification of Unmanned Aircraft Systems," Notice of Proposed Rulemaking, 84 Fed. Reg. 72438 (December 31, 2019).

30. "Remote Identification of Unmanned Aircraft Systems," Final Rule, 786 Fed. Reg. 4390 (January 15, 2021).

31. Standard Specification for Detect and Avoid System Performance Requirements, F3442/F3442M - 20 (May 2020).

32. "Type Certification of Unmanned Aircraft Systems," Notice of Policy, FAA-2019-1038, 85 Fed. Reg. 505 (February 3, 2020).

33. These proposed criteria were published in the Federal Register, with comment periods of December 21, 23, and 24, 2020.

34. See Section 2208 of Pub. L. No. 114-190 (July 15, 2016), and Sections 376–377 of Pub. L. No. 115-254 (October 5, 2018).

35. FAA, "Unmanned Aircraft Systems (UAS) Traffic Management (UTM) Concept of Operations, Version 2.0," Office of NextGen, March 2, 2021, https://

www.faa.gov/uas/research_development/traffic_management/media/UTM_ConOps_v2.pdf.

36. See Drone Advisory Committee, DAC Meeting, October 22, 2020 (virtual): response to task group #7 recommendations on UTM.

37. Recently, the Government Accountability Office issued an analysis of federal, state, tribal, and local government responsibilities, concluding that case law was unsettled both as to the extent of state and local authority over drones and as to how the law of torts, including trespass and aerial trespass, applies to drones. U.S. Government Accountability Office, *Unmanned Aircraft Systems: Current Jurisdictional, Property, and Privacy Legal Issues Regarding Commercial and Recreational Use of Drones*, B-330570 (September 16, 2020).

38. Several states have enacted laws prohibiting local jurisdictions from enacting drone ordinances or requiring the approval of the state department of transportation before doing so.

39. FAA, Office of Chief Counsel, State and Local Regulation of Unmanned Aircraft Systems (UAS) Fact Sheet (December 17, 2015), p. 3. The FAA issued this guidance in the wake of a flurry of state and local government efforts to regulate drones.

40. See FAA, Busting Myths about the FAA and Unmanned Aircraft (March 7, 2014). ("The FAA is responsible for the safety of U.S. airspace from the ground up.")

41. FAA Fact Sheet, p. 3.

42. FAA Fact Sheet, pp. 5–7.

43. 284 F. Supp. 3d at 132.

44. 49 U.S.C. § 41713(b)(1).

45. American Law Institute, *Restatement of the Law, Second, Torts* [hereinafter *Restatement*], vol. 1 (St. Paul, MN: American Law Institute Publishers, 1965), pp. 277–79.

46. 328 U.S. at 260.

47. *United States v. Causby*, 328 U.S. at 264.

48. 49 U.S.C. § 40102(a)(32) (emphasis added).

49. *Griggs*, 369 U.S. at 88–89.

50. *Restatement*, vol. 1, p. 281.

51. *Restatement*, vol. 1, pp. 283–84 (emphasis added).

52. The Drone Tort Committee determined that privacy and other torts did not warrant special consideration. National Conference of Commissioners on Uniform State Laws, "Uniform Tort Law Relating to Drones Act, Draft for Approval," May 30, 2019, https://www.uniformlaws.org/HigherLogic/System/DownloadDocumentFile.ashx?DocumentFileKey=5bbdd6ae-9c3f-7a80-6a0f-6cdf54dbfe9e&forceDialog=0.

53. National Conference of Commissioners on Uniform State Laws, "Drones Act, Draft," pp. 6–7.

54. National Conference of Commissioners on Uniform State Laws, "Drones Act, Draft," p. 7.

55. National Conference of Commissioners on Uniform State Laws, "Drones Act, Draft," p. 8.

56. American Law Institute, *Restatement of the Law, Second, Torts*, vol. 4 (1979), p. 101.

57. This chapter does not address state privacy laws, such as the California Consumer Privacy Act, which regulates the collection, retention, and use of personal identifiably information. This chapter also does not address the privacy interests protected by the Fourth Amendment to the U.S. Constitution. As yet, no federal privacy law applies to drone operations. Indeed, the FAA has proclaimed that it does not want to deal with the issue.

President Barack Obama by Executive Memorandum in February 2015 directed the National Telecommunications and Information Administration (NTIA) within the Commerce Department to deal with the privacy implications of drones. NTIA does not have regulatory authority. NTIA set up a multistakeholder process with the objective of reaching a consensus on best practices and published a set of best practices in May 2016. NTIA, "Voluntary Best Practices for UAS Privacy, Transparency, and Accountability," May 18, 2016, ntia.doc.gov/files/ntia/publications/uas_privacy_best_practices_6-21-16.pdf.

58. American Law Institute, *Restatement of the Law, Second, Torts*, vol. 3 (1977) § 652B, p. 378.

59. *Restatement*, vol. 3, p. 378.

CHAPTER THREE

1. Jackie Alkobi, *The Evolution of Drones: From Military to Hobby and Commercial* (blog), Percepto, January 15, 2019.

2. Alkobi, *Evolution of Drones*.

3. Alkobi, *Evolution of Drones*.

4. John Kelvey, "Drone Pilots Also Navigate Conflicting County/State Laws" (originally in *Carroll County Times*, Westminster, MD), February 7, 2020, Governing, https://www.governing.com/now/Drone-Pilots-Also-Navigate-Conflicting-County-State-Laws.html.

5. Rhodri Marsden, "Rhodri Marsden's Interesting Objects: Lord Winchelsea's Red Flag," *Independent.co*, July 3, 2015.

6. James Czerniawski and Nick Grose, "Utah Should Embrace the Innovation That Springs from Economic Disruption," *Salt Lake Tribune*, August 13, 2020.

7. Scott Brinker, "Martec's Law: Technology changes exponentially, organizations change logarithmically," Chiefmartec.com (blog), June 13, 2013.

8. CompTIA, "The Drone Market: Insights from Customers and Providers," Research Report, June 2019, https://www.comptia.org/content/research/drone-industry-trends-analysis#:~:text=IDCiii%20predicts%20that%20the,over%20year%20increase%20of%2013%25.

9. Michał Mazur and Adam Wiśniewski, *Clarity from Above: PwC Global Report on the Commercial Applications of Drone Technology*, PwC, May 2016.

10. Mazur and Wiśniewski, *Clarity from Above*, p. 18.

11. Mazur and Wiśniewski, *Clarity from Above*, p. 18.

12. Mazur and Wiśniewski, *Clarity from Above*, p. 18.

13. Mazur and Wiśniewski, *Clarity from Above*, p. 18.

14. Jeremy Jensen, "Agricultural Drones: How Drones Are Revolutionizing Agriculture and How to Break into This Booming Market," UAVCoach.com, April 18, 2019.

15. Benjamin Pinguet, "The Role of Drone Technology in Sustainable Agriculture," PrecisionAG.com, April 22, 2020.

16. Lutz Goedde et al., "Agriculture's Connected Future: How Technology Can Yield New Growth," McKinsey.com, October 9, 2020.

17. Jahnavi Sajip, "How Can Drones Be Used in Building Projects," *NY Engineers* (blog), November 18, 2019.

18. Zacc Dukowitz, "Drones in Construction: How Drones Are Helping Construction Companies Save Money, Improve Safety Conditions, and Keep Customers Happy," uavcoach.com, June 6, 2020.

19. Dukowitz, "Drones in Construction."

20. Karissa Rosenfield, "These Two Drones Just Built a Bridge," ArchDaily.com, September 13, 2015.

21. Lidija Grozdanic, "How Drones Can Be Used in Architecture," Archipreneur.com, December 7, 2017.

22. Elena Mazarenu, "Household Package Mail Volume in the United States from FY 2014 to FY 2019," Statista.com, June 5, 2020.

23. "Global Parcel Delivery Market Insight Report (2020–2024): Impact of the COVID-19 Pandemic," *GlobNewswire.com*, April 24, 2020.

24. Aaron Pressman, "Drone Industry Flies Higher as COVID-19 Fuels Demand for Remote Services," *Fortune.com*, July 13, 2020.

25. Charlie Rose, "Amazon's Jeff Bezos Looks to the Future," *60 Minutes*, CBS News, December 1, 2013.

26. David Pierce, "Delivery Drones Are Coming: Jeff Bezos Promises Half-Hour Shipping with Prime Air," *The Verge*, December 1, 2013.

27. "Amazon Makes First Drone Delivery," *BBC News*, December 14, 2016.

28. Luke Dormehl, "When It Comes to Delivery Drones, Google's Wing Is Miles above the Competition," *Digital Trends*, January 27, 2020.

29. Dormehl, "Google's Wing."

30. Akshat Sharma, "How Future Delivery Drones Will Deliver Your Packages," JungleWorks.com, June 20, 2019.

31. Anmar Frangoul, "Autonomous Drone Delivers Diabetes Medication to a Remote Irish Island," CNBC, September 18, 2019, https://www.cnbc.com/2019/09/18/autonomous-drone-delivers-diabetes-medication-to-a-remote-irish-island.html.

32. Felicia Shivakumar, "Giant Cargo Drones Will Deliver Packages Farther and Faster," *The Verge*, June 10, 2019.

33. Shivakumar, "Giant Cargo Drones."

34. Morgan Stanley Research, "Flying Cars and Autonomous Aircraft," MorganStanley.com, January 23, 2019.

35. Morgan Stanley Research, "Flying Cars."

36. Morgan Stanley Research, "Flying Cars."

37. Morgan Stanley Research, "Flying Cars."

38. Jessica Purkiss, and Jack Serle, "Obama's Covert Drone War in Numbers: Ten Times More Strikes Than Bush," The Bureau of Investigative Journalism, January 17, 2017.

39. Micah Zenko, *Obama's Final Drone Strike Data* (blog), Council on Foreign Relations, January 20, 2017.

40. "The Secret Death Toll of America's Drones," editorial, *New York Times*, March 30, 2019.

41. Stephen Rice, "10 Ways That Police Use Drones to Protect and Serve," *Forbes*, October 7, 2019.

42. Rice, "10 Ways That Police Use Drones."

43. Ben Yakas, "The NYPD Is Monitoring Social Distancing on the Ground and from the Skies," *Gothamist*, March 31, 2020.

44. Travis Fedschun, "Coronavirus Spurs Police to Deploy 'Talking' Drones in Florida, New Jersey to Enforce Social Distancing," Fox News, April 9, 2020.

45. Guest Blogger, *Drones Shine in Emergency Management* (blog), IBM Center for the Business of Government, February 11, 2019.

46. "Drones Have Saved the Lives of 279 People," Heliguy.com, October 7, 2019.

47. Chris Huber, "2017 Hurricane Maria Stats: Facts, FAQ, and How to Help," World Vision (updated), August 1, 2018.

48. Rachel Becker, "Trying to Communicate after the Hurricane: 'It's as If Puerto Rico Doesn't Exist'," *The Verge*, September 29, 2017.

49. Vann R. Newkirk II, "A Year after Hurricane Maria, Puerto Rico Finally Knows How Many People Died," *The Atlantic*, August 28, 2018.

50. Newkirk, "A Year after Hurricane Maria."

51. John Brodkin, "An AT&T Drone Is Now Providing Cellular Service to People in Puerto Rico," *Ars Technica*, November 6, 2017.

52. Liz Meszaros, "Drone Technology: A New Ally in the Fight against COVID-19," MDLinx, April 8, 2020.

53. Meszaros, "Drone Technology."

54. Stephanie Zacharek, "How Drones Are Revolutionizing the Way Film and Television Is Made," *Time*, May 31, 2018.

55. Zacharek, "How Drones Are Revolutionizing Film."

56. Zacharek, "How Drones Are Revolutionizing Film."

57. Federal Aviation Administration (FAA), "FAA Statement: Federal vs. Local Drone Authority," press release, July 20, 2018. See 49 U.S.C. § 40103(b)(1), on sovereignty and use of airspace.

58. FAA, "Unmanned Aircraft Systems Operations in the U.S. National Airspace System," 72 Fed. Reg. 6689 (Feburary 13, 2007).

59. FAA, "Unmanned Aircraft Systems Operations."

60. Small Unmanned Aircraft Systems, 14 C.F.R. §§ 107.1, 107.3 (2019).

61. Small Unmanned Aircraft Systems.

62. Small Unmanned Aircraft Systems.

63. H.R. 2997, 21st Century AIRR Act, 115th Cong., June 22, 2017, https://www.congress.gov/bill/115th-congress/house-bill/2997.

64. Brianna Fernandez, "Is It Time for the U.S. to Privatize Its Air Traffic Control?," American Action Forum, September 12, 2017.

65. Fernandez, "Is It Time?"

66. Glenn Farley, "Canada ahead of the US with Some Air Traffic Control Technologies," *King5.com*, November 27, 2019.

67. Adam D. Thierer, *Permissionless Innovation: The Continuing Case for Comprehensive Technological Freedom*, 2nd ed. (Arlington, VA: Mercatus Center at George Mason University, 2016), p. 9.

68. Libertas Institute, Permissionless Innovation, https://freemarketutah.com/.

69. Adam Vaughan, "McDonald's to Switch to Paper Straws in UK after Customer Campaign," *The Guardian*, June 15, 2018, https://www.theguardian.com/business/2018/jun/15/mcdonalds-to-switch-to-paper-straws-in-uk-after-customer-concern.

70. Cass R. Sunstein, *Laws of Fear: Beyond the Precautionary Principle* (Cambridge, UK: Cambridge University Press, 2005), p. 4.

71. Small Unmanned Aircraft Systems, 14 C.F.R. §§ 107.1, 107.31 (2019).

72. Small Unmanned Aircraft Systems.

73. Dave Lee, "DJI Makes App to Identify Drones and Find Pilots," BBC, November 14, 2019.

74. Stan Horaczek, "DJI's Mavic Air Drone Uses More Than a Dozen Sensors to Keep It from Crashing," *Popular Science*, January 23, 2018.

75. Horaczek, "DJI's Mavic Air Drone."

76. See Maryland Robotics Center, "UMD Risk Expert Contributes to National Academies Report on National Drone Policy," University of Maryland, June 13, 2018.

77. Maryland Robotics Center, "UMD Risk Expert."

78. Pub. L. No. 115-254 (October 5, 2018).

79. Brent Skorup and Connor Haaland, "Which States Are Prepared for the Drone Industry?," *Mercatus Research*, March 19, 2020.

80. Skorup and Haaland, "Which States Are Prepared?"

81. Skorup and Haaland, "Which States Are Prepared?"

82. Skorup and Haaland, "Which States Are Prepared?"

83. When considering government use of air highways, it needs to be tied to the highway's intended use. For example, a state entity should not be able to use an air highway to conduct surveillance except in limited and clearly defined circumstances. Using the highway in explicitly different ways than the intended commercial use requires careful consideration, and those proposed uses need to be treated appropriately

84. Air right, Dictionary.com, https://www.dictionary.com/browse/air-right.

85. Skorup and Haaland, "Which States Are Prepared?"

86. Brent Skorup, "Drone Technology, Airspace Design, and Aerial Law in States and Cities," Mercatus Working Paper, December 2, 2020.

87. Skorup and Haaland, "Which States Are Prepared?"

88. Max Pedowicz, "Managing the Airspace: Maximizing the Easement Opportunities as Drones Proliferate," *Right of Way*, May/June 2016.

89. Pedowicz, "Managing the Airspace."

90. Cailin Crowe, "Surveillance Planes to Fly over Baltimore amid COVID-19," Smart Cities Dive, April 3, 2020.

91. While the Baltimore case is not a situation of drone use by law enforcement, as mentioned in the use case section, police have used drones for surveillance purposes. Americans need to have this conversation to determine the level of regulation that citizens think is appropriate for that intended use.

92. H.B. 243, 2021 Gen. Sess. (Utah, 2021), https://le.utah.gov/~2021/bills/static/HB0243.html.

93. Libertas Institute Staff, "Proposal: The Privacy Protection Act," Libertas Institute, May 14, 2020. The legislation was drafted in response to events surrounding a Utah-based company known as Banjo, which utilized artificial intelligence technology and connected with state surveillance infrastructure (closed circuit television, traffic cameras, 911 phone calls, etc.) to improve response time by emergency service members. After the revelation that the founder and chief executive of the company had former ties to the Ku Klux Klan, the state suspended its

existing contracts with the company and is auditing the technology for any potential bias against race. Many Utahns were unaware of the arrangement the company had with the state and were upset upon its discovery. There was bipartisan desire to achieve meaningful change, and the Privacy Protection Act proposal received widespread support across the political spectrum. However, the language is not finalized, with the bill being considered in Utah's 2020 general session and subject to change after discussions with various stakeholders.

94. "Justice Alito: Legislatures Must Pass 21st Century Privacy Laws, Can't Be Left to Courts," *Privacy SOS* (blog), ACLUm, September 21, 2015, https://privacysos.org/blog/justice-alito-legislatures-must-pass-21st-century-privacy-laws-cant-be-left-to-courts/.

95. FAA, UAS Integration Pilot Program, https://www.faa.gov/uas/programs_partnerships/integration_pilot_program/.

96. FAA, "BEYOND," https://www.faa.gov/uas/programs_partnerships/beyond/.

97. Ryoji Kashiwagi, "The Rise of the Regulatory Sandbox," *Financial IT*, April 12, 2017.

98. Molly Jane Zuckerman, "UK Financial Regulator Introduces Global Fintech Sandbox, '90%' Success Rate Domestically," Cointelegraph, March 19, 2018.

99. Out of concern for public health and safety, the government requires some companies to buy certain kinds of insurance and reserves the right to cancel a company's participation in the program altogether.

100. Aaron Stanley, "Arizona Becomes First U.S. State to Launch Regulatory Sandbox for Fintech," *Forbes*, March 23, 2018.

101. Bob Ambrogi, "Utah Supreme Court Votes to Approve Pilot Allowing Non-Traditional Legal Services," *LawSites* (blog), August 29, 2019.

102. Lyle Moran, "Utah Embraces Nonlawyer Ownership of Law Firms as Part of Broad Access-to-Justice Reforms," *ABA Journal*, August 14, 2020.

103. Lyle Moran, "Rocket Lawyer Is among the First Applicants Approved to Join Utah's Regulatory Sandbox Program," *ABA Journal*, September 8, 2020.

104. FAA, UAS Integration Pilot Program.

105. FAA, UAS Integration Pilot Program.

106. FAA, UAS Integration Pilot Program.

107. John Nancarrow, "CFPB Sandbox Outlook, Applying FCRA to Feds, Calif. Treasurer," Bloomberg Law, December 28, 2018.

108. *Merriam-Webster,* s.v. "crony capitalism," https://www.merriam-webster.com/dictionary/crony%20capitalism.

109. Arizona Attorney General Mark Brnovich, "Sandbox Participants," https://www.azag.gov/fintech/participants.

110. Office of Legal Services Innovation, "Innovation Office Interim Update Report," Utah Supreme Court, November 20, 2020.

111. Elizabeth Rossiello, "Forget Sandboxes: Why Governments Should Accept Licensing from Other Jurisdictions," *Forbes*, February 14, 2019.

112. Annie Palmer, "Amazon Wins FAA Approval for Prime Air Drone Delivery Fleet," CNBC.com, August 31, 2020.

CHAPTER FOUR

1. As economist Tyler Cowen noted in a January 2020 interview about the obstacles to drone delivery, "it's a regulatory issue," https://www.facebook.com/fitzgerald.brendan.wallace/posts/10101719850747122.

2. Brendan Fitzgerald Wallace Facebook page, "My conversation with economist, author & podcaster Tyler Cowen," January 25, 2020, https://www.facebook.com/fitzgerald.brendan.wallace/posts/10101719850747122.

3. Evan Ackerman, "Zipline Wants to Bring Medical Drone Delivery to U.S. to Fight COVID-19," *IEEE Spectrum*, April 3, 2020.

4. Miriam McNabb, "UPS Drone Delivery: DroneUp Flies to Prove the Case for Coronavirus Response," Dronelife, April 21, 2020.

5. Federal Aviation Administration (FAA), "Busting Myths about the FAA and Unmanned Aircraft," last modified March 7, 2014.

6. Andy Pasztor, "FAA Projects Fourfold Increase in Commercial Drones by 2022," *Wall Street Journal*, March 18, 2018.

7. FAA, *FAA Strategic Plan for FY 2019–2022*, February 2018.

8. FAA, "UAS by the Numbers," last modified March 1, 2021.

9. See Ryan Hagemann, Jennifer Huddleston Skees, and Adam Thierer, "Soft Law for Hard Problems: The Governance of Emerging Technologies in an Uncertain Future," *Colorado Technical Law Journal* 17.1 (2018): 37.

10. Uber, "Day 1: Uber Elevate Summit 2019 | Uber," June 11, 2019, YouTube video, 9:11:05, https://www.youtube.com/watch?v=YMBEev4MmU4, at 2:43.

11. Del. Code tit. 11, § 1334 (no operations over events with more than 1,500 people, over incidents with first responders, or over critical infrastructures); Fla. Stat. Ann. § 330.41 (no operations over critical infrastructures); La. Stat. Ann. § 14:337 (no operations over certain facilities, including schools); La. Stat. Ann. § 14:63 (defining criminal trespass to include operating a UAS over property of another); La. Stat. Ann. § 14:108 (no operations over a police cordon); Nev. Rev. Stat. § 493.103 (no operations under 250 feet over property when owner notifies UAS operator); Nev. Rev. Stat. § 493.109 (no operations over new critical facilities); Okla. Stat. tit. 3, § 322 (no operations over critical infrastructure facilities);

Or. Rev. Stat. § 837.380 (restrictions over property when owner notifies UAS operator); S.D. Codified Laws § 50-15-3 (no operations over prisons or military facilities); Tenn. Code Ann. § 39-14-405 (criminal trespass via UAS over private property in nonnavigable airspace); Tex. Gov't Code Ann. §§ 411.062, 411.065 (restrictions over state Capitol Complex); Tex. Gov't Code Ann. § 423.0046 (restrictions over sports venues); Wis. Stat. § 114.045 (no operations over a correctional institution). See Stephen J. Migala, "UAS: Understanding the Airspace of States," *Journal of Air Law and Commerce* 82, no. 1 (2017): 62–63.

12. In October 2019, Silverthorne, Colorado, passed an ordinance that generally prohibits drones at heights less than 40 feet above rooftops. See John Minor, Chief of Police, to Town Council, Ordinance 2019-19, An Ordinance to Provide for the Regulation of Unmanned Aircraft, Town of Silverthorne Council Agenda Memorandum, November 13, 2019.

13. Alissa M. Dolan and Richard M. Thompson II, "Integration of Drones into Domestic Airspace: Selected Legal Issues," Congressional Research Service Report R42940, April 4, 2013.

14. See, for example, Brief of Amicus Curiae Association for Unmanned Vehicle Systems and the Consumer Technology Association 7, *National Press Photographers v. McCraw*, Civ. Action No. 1:19-cv-00946-RP, January 10, 2020 (asserting that state laws cannot "directly limit the operations of UAS in the national airspace" by creating no-fly zones above sensitive locations).

15. Betsy Lillian, "Drone Industry Responds to Draft Tort Law on 'Aerial Trespass,'" Unmanned Aerial, July 25, 2018 (quotation attributed to representatives in the drone industry).

16. RTCA, "Drone Integration Funding," Report of the Drone Advisory Committee, RTCA Paper No. 047-18/DAC-011, March 2018.

17. U.S. Government Accountability Office, "Unmanned Aircraft Systems: FAA Should Improve Drone-Related Cost Information and Consider Options to Recover Costs," GAO-20-116, December 2019, p. 38; U.S. Government Accountability Office, "Unmanned Aircraft Systems: Current Jurisdictional, Property, and Privacy Legal Issues Regarding the Commercial and Recreational Use of Drones," Appendices I–VI, B-330570, September 2020.

18. David Manheim, John F. Raffensperger, and Jia Xu, "Regulating Drone Airspace Using 'Smart Markets,'" *The RAND Blog*, RAND Corporation, April 19, 2016.

19. Antony D. Evans, Maxim Egorov, and Steven Munn, "Fairness in Decentralized Strategic Deconfliction in UTM," AIAA Scitech 2020 Forum, American Institute of Aeronautics and Astronautics, January 2020, https://storage.googleapis.com/blueprint/Fairness_in_Decentralized_Strategic_Deconfliction_in_UTM.pdf.

20. Brent Skorup, "Auctioning Airspace," *North Carolina Journal of Law and Technology* 21, no. 1 (2019): 79 (proposing the demarcation and auction of aerial corridors for passenger drones).

21. See Virginia K. Trunkes, "Balancing New Technology and Privacy When Using Drones in Land Use and Construction," *National Law Review* X, no. 147 (May 26, 2020) ("In 2019, the American Law Institute's . . . drafters of the Fourth Restatement of Property applied principles of trespass law in proposing § 1.2A—'Trespass by Overflight.'"); Tort Law Related to Drones Act § 301(a), Proposed Draft for Discussion, National Conference of Commissioners on Uniform State Laws (2018).

22. The famous *ad coelum* maxim is this: "Land hath also, in its legal signification, an indefinite extent, upwards as well as downwards." William Blackstone, *Commentaries on the Laws of England in Four Books*, vol. 2 (1818), p. 18. The Supreme Court in *United States v. Causby* somewhat exaggerated "*indefinite* extent" in the original maxim to mean something like "*infinite* extent." *United States v. Causby*, 328 U.S. 256, 260 (1946) (characterizing *ad coelum* as the "ancient doctrine that at common law ownership of the land extended to the periphery of the universe"). Courts cited the maxim frequently in trespass cases but typically denied that there could be a trespass or a property interest in airspace that was not practicably usable by the landowner. See, for example, *Johnson v. Curtiss N. W. Airplane Co.*, 1928 U.S. Av. R. 42, 43–44 (Dist. Ct., Ramsey Co., Minn., 1923) ("[W]hen, as here, the air is to be considered at an altitude of two thousand feet or more, to contend that it is part of the realty . . . is only a legal fiction, devoid of substantial merit."). Possibly the first Anglo-American statute recognizing landowners' rights to surface airspace is the Telegraph Act of 1863 in the United Kingdom, which gave landowners a right to object to the construction of a telegraph line above their property. Telegraph Act of 1863, c. 112 sec. 22 (specifying that "the [telegraph] Company shall not place a Telegraph above Ground . . . or place a Telegraph above Ground across an Avenue or Approach to a Dwelling House" unless "in each Case [the company] obtain the consent of the" occupier, lessee, or owner). See also John Frederic Clerk and W. H. B. Lindsell, *The Law of Torts*, 2nd ed. (London: Sweet & Maxwell, 1896), p. 291 ("The provisions of the Telegraph Act, 1863 . . . , are based upon the assumption that there is a right of property in the air space." . . .) (citing Telegraph Act of 1863, c. 112).

23. A New York court in 1906, for instance, permitted the ejectment of a telephone line above property but warned that "this [*ad coelum* maxim] may not be taken too literally." *Butler v. Frontier Telephone Co.*, 186 NY 486, 491, 79 NE 716, 718 (1906).

24. John B. Phear, *A Treatise on the Rights of Water* (London: V. & R. Stevens and G.S. Norton, 1859), p. 2 ("[T]he partition [of land] may be carried on in a vertical, as well as in a horizontal direction.").

25. "[I]t is curious to note that even as late as the early part of the last [that is, 19th] century, there was considerable doubt as to whether trespass would lie, where there was no tangible interference with the land, but only with the airspace." Arthur K. Kuhn, "The Beginnings of an Aërial Law," *American Journal of International Law* 4, no. 1 (1910): 109, 123. See also Silas Jones, *An Introduction to Legal Science* (New York: J.S. Voorhies, 1842), p. 179 (noting exceptions to *ad coelum*: "[F]or instance, a man may have an inheritable corporeal property in an upper chamber, though the lower stories and soil may belong to another. This, it is true, is as much as saying a man may have land by owning an upper chamber, or in other words, that an upper chamber is land!"); Joseph A. Shearwood, *A Concise Abridgment of the Law of Real Property and an Introduction to Conveyancing* (London: Stevens & Sons, 1878), p. 2 ("One man therefore may have a house in fee and another the ground in fee; or if the house is subdivided in chambers there may be different owners in fee to each set."). As one treatise noted:

> The English law is different [from the absolute ownership principles in Roman law], permitting one man to own the surface, another to own a mining substratum, while still a third owns a horizontal flat in the structure erected upon the land. Accordingly, I say, the adoption of a zone theory would be quite in harmony with the general spirit of the English land law as regards these horizontal hereditaments.

Harold D. Hazeltine, *The Law of the Air* (London: University of London Press, 1911), p. 75.

26. John G. Hawley and Malcolm McGregor, *A Treatise on the Law of Real Property* (Detroit, MI: Collector Publishing Co., 1900), p. 5.

27. See Theodore Stein, *Slide Mountain: Or, the Folly of Owning Nature* (Oakland, CA: University of California Press, 1995), pp. 141–46 (noting that after passage of the 1916 law, "airspace, a three-dimensional abstraction, became a *thing* that could be owned and sold.").

28. Stein, *Slide Mountain*, pp. 147–48 ("Transferring air was nothing new in New York. The city had permitted the shifting of air rights from lot to lot to build higher towers at various times since the 1920s.").

29. Theodore Schmidt, "Public Utility Air Rights," *Journal of Air Law and Commerce* 1, no. 1 (1930): 52, 68.

30. Schmidt, "Public Utility Air Rights."

31. Schmidt, "Public Utility Air Rights," p. 68. A 1929 conveyance of air rights in Boston was done using common law principles. Schmidt, "Public Utility Air Rights," pp. 70–71.

32. Schmidt, "Public Utility Air Rights," pp. 53–54 (quoting former American Bar Association president).

33. See Douglas C. Harris, "Condominium and the City: The Rise of Property in Vancouver," *Law and Social Inquiry* 36, no. 3 (2011): 694, 695.

34. See "Conveyance and Taxation of Air Rights," *Columbia Law Review* 64, no. 2 (1964): 338, 339 ("[I]t seems reasonably clear that an owner may effectively convey 'tracts' of space that are physically unattached to the land."). The American Bar Association published the Model Airspace Act in 1972, which formalized the process of granting property rights for airspace, but it appears that only Oklahoma adopted significant portions of the act. See Subcommittee on Airspace Utilization and Multiple Use, Committee on New Developments in Real Estate Practice, "Final Draft of Model Airspace Act," *Real Property, Probate and Trust Journal* 7, no. 2 (1972): 353–81; 60 Okl. St. § 802 et seq.

35. See, for example, Va. Code Ann. § 55.1-1900 (2020) (defining "land" as "a three-dimensional concept"); *Macht v. Department of Assessments*, 266 Md. 602, 612–13 (1972) (holding that for assessment purposes, airspace is treated like a negative easement for a term of years); 68 P.S. § 802 (2020) (stating that "rights and interests in air space . . . shall be dealt with for all purposes and in all respects as estates, rights and interests in real property"); In re. Appeal of Bigman, 110 Pa. Commw. 539, 547 (1987) ("Although air space does not fit squarely within either of these definitions, we conclude that it is more closely aligned with 'buildings' than with 'land.'"). But see *Penn Central Transportation v. City of New York*, 438 U.S. 104 (1978) (holding that air rights do not constitute real property in the context of regulatory takings).

36. Hazeltine, *The Law of the Air*, pp. 15, 46–47 (noting that air above land is "an appurtenance of the territorial state or even . . . a part of the territorial state"). Though Hazeltine is speaking of nations when referring to "states," in the United States land and territory are generally held and regulated by subnational states upon joining the union.

37. Section 2 of the 1922 Uniform State Law for Aeronautics provided, "Sovereignty in the space above the lands and waters of this State is declared to rest in the State, except where granted to and assumed by the United States pursuant to a constitutional grant from the people of this State."

38. Ariz. Rev. Stat. § 28-8206 (2020); Cal. Pub. Util. Code § 21401 (2020); C.R.S. 41-1-106 (2020); 2 Del. C. § 302 (2020); Haw. Rev. Stat. Ann. § 263-2 (2020); Idaho Code § 21-202 (2020); Burns Ind. Code Ann. § 8-21-4-2 (2020); 1 M.R.S. § 6 (2020); Md. Transp. Code Ann. § 5-104 (2020); Minn. Stat. § 360.012 (2020); Mont. Code Ann. 67-1-202 (2019); Nev. Rev. Stat. Ann. § 493.030 (2019); N.C. Gen. Stat. § 63-11 (2020); N.D. Cent. Code § 2-03-02 (2019); N.J. Stat. § 6:2-4 (2020); Tenn. Code Ann. § 42-1-102 (2020); Utah Code Ann. § 72-10-123 (2020); Wis. Stat. § 114.02 (2020); Wyo. Stat. § 10-4-301 (2020). South Carolina, South Dakota, and Vermont repealed their airspace sovereignty statutes in 2012,

2014, and 1997, respectively. See S.C. Code Ann. § 55-3-30 (repealed 2012); S.D. Codified Laws § 50-13-2 (repealed 2014); 5 V.S.A. § 401 (repealed 1997).

39. 44 Stat. 568, Air Commerce Act of 1926, Sec. 6 (1926). In 1938, this declaration was amended somewhat, though it was still interpreted to mean sovereignty against foreign nations. See Laura K. Donohue, "A Tale of Two Sovereigns: Federal and State Use and Regulation of Unmanned Aircraft Systems," in *Handbook of Unmanned Aerial Vehicles*, 2nd ed., ed. Kimon P. Valavanis and George J. Vachtsevanos (New York: Springer International Publishing AG, 2017), pp. 35–36.

40. See, for example, Lake Tahoe Airport, "Who Controls and Enforces How Aircraft and Drones Operate in the Airspace above Lake Tahoe?," City of South Lake Tahoe, CA, July 24, 2017 ("The United States Government has exclusive sovereignty over airspace of the United States pursuant to 49 U.S.C.A. § 40103. The airspace, therefore, is not subject to private ownership."); Cyrus Farivar, "After Neighbor Shot Down His Drone, Kentucky Man Files Federal Lawsuit," *Ars Technica*, January 6, 2016.

41. Sen. Hiram Bingham, one of the drafters of the law, confirmed that the act made "no interference with municipal or State regulation." 67 Cong. Record 9355 (1926) (statement of Senator Bingham) (stating, "None whatever.").

42. In his influential legal brief to the U.S. Senate about the 1926 act, Senate legislative counsel Frederic P. Lee noted that the sovereignty provisions left surface air rights unaffected:

> It is true that the principle of exclusive Federal sovereignty in the air domain *above the surface air space*, rests the validity of such diverse State regulations (so far as they apply to the upper strata of air space) only upon the consent of the Federal Government rather than upon a State power which may be exercised irrespective of the action of the Federal Government.

Frederic P. Lee, "The Air Domain of the United States," in *Civil Aeronautics: Legislative History of the Air Commerce Act of 1926 and Revision of Law Memoranda upon Civil Aeronautics* (Washington: Government Printing Office, 1943), p. 131 (emphasis added).

43. The Court in *Braniff Airways* rejected the claim that the sovereignty provision nationalized airspace against the states:

> The provision pertinent to sovereignty over the navigable air space in the Air Commerce Act of 1926 was an assertion of exclusive national sovereignty. The convention between the United States and other nations respecting international civil aviation . . . accords. The Act, however, did not expressly exclude the sovereign powers of the states. . . . These

> Federal Acts regulating air commerce are bottomed on the commerce power of Congress, not on national ownership of the navigable air space, as distinguished from sovereignty.

Braniff Airways v. Neb. State Board of Equalization & Assessment, 347 U.S. 590, 595–97 (1954).

44. *United States v. Causby*, 328 U.S. 256 (1946).

45. *United States v. Causby*, 328 U.S. at 260 ("It is therefore argued [by the federal government] that, since these flights were within the minimum safe altitudes of flight which had been prescribed, they were an exercise of the declared right of travel through the airspace. The United States concludes that, when flights are made within the navigable airspace without any physical invasion of the property of the landowners, there has been no taking of property."). "None of [the Justices] cared that federal law had defined as navigable airspace the area in which the planes flew over the Causbys' land." Stuart Banner, *Who Owns the Sky?: The Struggle to Control Airspace from the Wright Brothers On* (Cambridge, MA: Harvard University Press, 2008), p. 250.

46. *United States v. Causby*, 328 U.S. at 260 ("[The United States] also argues that the landowner does not own superadjacent airspace which he has not subjected to possession by the erection of structures or other occupancy.").

47. *United States v. Causby*, 328 U.S. at 264–65.

48. *United States v. Causby*, 328 U.S. at 264.

49. *United States v. Causby*, 328 U.S. at 266. Sovereignty in the airspace rests in the state "except where granted to and assumed by the United States." Gen. Stats. 1943, § 63-11.

50. *United States v. Causby*, 328 U.S. at 266 (quoting United States ex rel. *TVA v. Powelson*, 319 U.S. 266 (1943)). See also *Stop the Beach Renourishment Inc. v. Florida Department of Environmental Protection*, 130 S. Ct. 2592, 2597 (2010) ("*Generally speaking, state law defines property interests*, including property rights in navigable waters and the lands underneath them.") (internal citation omitted, emphasis added); *Board of Regents of State Colleges v. Roth*, 408 U.S. 564, 577 (1972).

51. The Supreme Court said that the Takings Clause "only protects property rights as they are established under state law, not as they might have been established or ought to have been established." *Stop the Beach Renourishment Inc.*, 130 S. Ct. at 2612.

52. *Griggs v. Allegheny County*, 369 U.S. 84, 88–89 (1962). Navigable airspace in this case—the takeoff and landing airspace—was within 11 feet of the homeowner's house, and aircraft were as close as 30 feet from the house.

53. Colin Cahoon, "Low Altitude Airspace: A Property Rights No-Man's Land," *Journal of Air Law and Commerce* 56, no. 1 (1990): 157, 191. See *McCarran International Airport v. Sisolak*, 127 S. Ct. 1260 (2007).

54. Gareth R. Jones and Michael W. Pustay, "Interorganizational Coordination in the Airline Industry, 1925–1938: A Transaction Cost Approach," *Journal of Management* 14 (1988): 529, 537.

55. Rich Freeman, "Walter Folger Brown: The Postmaster General Who Built the U.S. Airline Industry," U.S. Centennial of Flight Commission.

56. Jones and Pustay, "Interorganizational Coordination," p. 537.

57. Paul M. Godehn and Frank E. Quindry, "Air Mail Contract Cancellations of 1934 and Resulting Litigation," *Journal of Air Law and Commerce* 21 (1954): 253, 261–62.

58. Godehn and Quindry, "Air Mail Contract Cancellations of 1934," p. 261.

59. Godehn and Quindry, "Air Mail Contract Cancellations of 1934," pp. 262–63.

60. Jones and Pustay, "Interorganizational Coordination," pp. 537–38.

61. See Freeman, "Walter Folger Brown: The Postmaster General."

62. Jones and Pustay, "Interorganizational Coordination," p. 538.

63. Jones and Pustay, "Interorganizational Coordination," p. 538.

64. Godehn and Quindry, "Air Mail Contract Cancellations of 1934," pp. 262–63.

65. See *Pacific Air Transport v. U.S.*, Ct. Cl. 649, 764–65 (1942).

66. Godehn and Quindry, "Air Mail Contract Cancellations of 1934," pp. 264–65.

67. Peter G. Klein, "Government as the Source of Monopoly: U.S. Airlines Edition," Mises Wire, Mises Institute, January 6, 2017.

68. Jones and Pustay, "Interorganizational Coordination," p. 539.

69. John T. Correll, "The Air Mail Fiasco," *Air Force Magazine*, March 1, 2008.

70. Correll, "The Air Mail Fiasco."

71. Jones and Pustay, "Interorganizational Coordination," p. 539; and Correll, "The Air Mail Fiasco."

72. Jones and Pustay, "Interorganizational Coordination," p. 540.

73. Brent Skorup, "Auctioning Airspace," *North Carolina Journal of Law and Technology* 21, no. 1 (2019): 79.

74. U.S. Government Accountability Office, "Federal Aviation Administration—Authority to Auction Airport Arrival and Departure Slots and to Retain and Use Auction Proceeds," B-316796, September 30, 2008.

75. Jerry Limone, "JetBlue, WestJet Win Airport Slots at LaGuardia and Reagan National," *Travel Weekly*, December 1, 2011.

76. Doug Cameron, "Surging Air Traffic Prompts Rethink of Airport Gate Rights," *Wall Street Journal*, July 6, 2019.

77. Clifford Winston, "How the Private Sector Can Improve Public Transportation Infrastructure," Mercatus Center Working Paper no. 14-16, George Mason University, June 2014.

78. Winston, "How the Private Sector Can Improve."

79. Winston, "How the Private Sector Can Improve."

80. John Palfrey and Urs Gasser, *Interop: The Promise and Perils of Highly Interconnected Systems* (New York: Basic Books, 2012), p. 107.

81. Palfrey and Gasser, *Interop*, pp. 43–44. Technology exists to reduce separation times, but because of the free-rider problem, airlines have declined to make the necessary equipment installations. Michael O. Ball, George L. Donohue, and Karla Hoffman, "Auctions for the Safe, Efficient, and Equitable Allocation of Airspace System Resources," in *Combinatorial Auctions*, ed. Peter Cramton, Yoav Shoham, and Richard Steinberg (Cambridge, MA: MIT Press, 2005), p. 960.

82. Palfrey and Gasser, *Interop*, p. 107.

83. Palfrey and Gasser, *Interop*, p. 261.

84. Zach Wichter, "Inside the Room Where the FAA Controls U.S. Airspace," The Points Guy, October 23, 2019.

85. Palfrey and Gasser, *Interop*, p. 107.

86. Michael S. Greve, "Bloc Party Federalism," *Harvard Journal of Law and Public Policy* 42 (2019): 279, 287 (noting that cooperative federalism "works where and when states are tolerably homogenous" and "breaks down when a substantial number of states, *acting as a bloc*, refuse cooperation") (emphasis in original).

87. See Brent Skorup and Connor Haaland, "Which States Are Prepared for the Drone Industry? A Fifty-State Report Card," Mercatus Research Paper, Mercatus Center at George Mason University, March 19, 2020.

88. Federal-Aid Highway Program of 1961, Sec. 104. Codified at 23 U.S.C. 111.

89. Stephen S. Roop and Sondip Mathur, "Leasing of TxDOT's Rights-of-Way," Research Report 1329-1F, Texas Transportation Institute at Texas A&M University (1993), p. 2.

90. Roop and Mathur, "Leasing of TxDOT's Rights-of-Way."

91. Leo Thompson, "Is Your City Infrastructurally Obese?," Strong Towns, October 29, 2019.

92. Thompson, "Is Your City Infrastructurally Obese?"

93. Office of Transportation Policy Studies, "Future Uses of Highway Rights of Way," Report Summary, Federal Highway Administration, April 2012.

94. Carson Poe and Gina Filosa, "Alternative Uses of Highway Rights-of-Way: Accommodating Renewable Energy Technologies," *Transportation Research Record: Journal of the Transportation Research Board* 2270, no. 1 (2012): 23–30.

95. See Brian Garrett-Glaser, "Talking UTM with Amit Ganjoo, Founder of ANRA Technologies," Aviation Today, February 19, 2020.

96. See J. Scott Hamilton, "Allocation of Airspace as a Scarce Natural Resource," *Transportation Law Journal* 22 (1994): 251, 262.

97. Hamilton, "Allocation of Airspace."

98. See Hamilton, "Allocation of Airspace."

99. Ricarda L. Bennett, "Airport Noise Litigation: Case Law Review," *Journal of Air Law and Commerce* 47 (1982): 449, 490. See, for example, *Greater Westchester Homeowners Association v. Los Angeles*, 160 Cal. Rptr. 733 (1979), cert. denied, 449 U.S. 820 (1980).

100. Several states that have considered aviation lawsuits have recognized prescriptive easements—essentially adverse possession of airspace. Courts in California, Oregon, Washington, and Connecticut recognize prescriptive easements of airspace. *Baker v. Burbank-Glendale-Pasadena Airport Authority*, 220 Cal. App. 3d 1602, 1609 (1990); *Insitoris v. City of Los Angeles*, 210 Cal. App. 3d 10, 14 (1989); *Christie v. Miller*, 719 P.2d 68, 70 (Or. Ct. App. 1986); *Peterson v. Port of Seattle*, 618 P.2d 67, 70 (Wash. 1980) (acknowledging that air navigation—avigation—easements for public use can be prescriptively acquired and are not compensable); *Ventres v. Godspeed Airport LLC*, 881 A.2d 937, 949 (Conn. 2005) (holding that airports can acquire prescriptive easement, including the right to enter neighboring land and cut trees). While not expressly acknowledging a prescriptive easement, a New York court similarly prevented development in an aviation case:

> Although the operations of the airport have expanded considerably since 1962, the claimant purchased the property with knowledge of the presence of an airport, and therefore assumed the risk of fluctuations in market value that might be caused by the existence of a nearby airport. In this case, it cannot be said that the claimant ever had a reasonable expectation that the building could be vertically expanded. As the operations of the airport increased, the possibility of expansion diminished. This was not a result of the taking but of the risk the property owner assumed upon purchase of the property.

3775 Genesee St. Inc. v. State, 415 NYS 2d 575, 585 (1979).

101. One complication under current law is that the 1978 act preempts state regulation of routes, prices, and services of "air carriers." 49 U.S.C. § 41713. In 2019, the FAA started granting air carrier certificates to some drone companies. See Jamie Freed, "FAA to Award First Drone Airline Licence in the Next Month," *Reuters*, April 10, 2019.

102. 47 U.S.C. § 253 (a)–(c).

103. FAA, *Urban Air Mobility Concept of Operations v1.0*, June 26, 2020.

104. For passenger drones and large freight drones, this federal-market model would likely work fairly well. Passenger drones and freight drones will fly at hundreds or thousands of feet above the ground, much like small aircraft and helicopters—where federal interests predominate. The construction of new private passenger droneports will raise zoning and property rights issues in new locations,

but the existing framework of federal airspace regulation and local zoning regulation applies in a straightforward manner to large drones given their commonalities with traditional aviation.

105. Robin Riedel and Shivika Sahdev, "Taxiing for Takeoff: The Flying Cab in Your Future," McKinsey & Company, January 2019, https://web.archive.org/web/20190109164929/https:/www.mckinsey.com/industries/travel-transport-and-logistics/our-insights/taxiing-for-takeoff-the-flying-cab-in-your-future.

106. UTM operator leadership may lead to UTM and drone companies "possibly even carving up airspace among themselves." James Poss, "An Open Letter to the New FAA Administrator, on Drones," Inside Unmanned Systems, September 6, 2019. Typically, "air navigation service providers have teamed up with UTM vendors. . . . But this kind of setup creates 'a natural monopoly.'" Paul Willis, "Crowded Skies," *The Future of Transportation: Executive Briefing 2019* (Surrey, UK: UKi Media & Events, 2019), p. 70, www.ukimediaevents.com/publication/ccded72e/72 (quoting Jonas Stjernberg). "Most experts agree that setting up a UTM is the starting point of drone integration." Willis, "Crowded Skies," p. 71.

107. See Nick Zazulia, "Why Automation Shouldn't Push Pilots Out of Air Taxis," *Aviation Today*, May 23, 2019 ("'Look at [drones], the UAS industry has struggled with detect-and-avoid for 20 years,' said fellow team-member Ken Goodrich, a senior research engineer who also advocates initially piloted operations.").

CHAPTER FIVE

1. For further discussion of the several forms that the maxim has taken over time, see D. E. Smith, "The Origins of Trespass to Airspace and the Maxim "Cujus Est Solum Ejus Est Usque Ad Coelum," *Trent Law Journal* 6 (1982): 33, 36.

2. See Titus Lucretius Carus, *De Rerum Natura*, 1.9 (referring to light in the sky: *placatumque nitet diffuso lumine caelum*); 6.650 (referring to the infinite universe and one sky, *caelum unum*); 6.1119–24 (considering the movement of air as bringing disease and destruction: e.g., *fit quoque ut, in nostrum cum venit denique caelum, / corrumpat reddatque sui simile atque allienum*); 6.8 (*divolgata vetus iam ad caelum gloria fertur*); 6.669 (*flamescere caelum*). Also note that in 350 BCE Greek philosopher Aristotle laid out the locomotion of the heavens and heavenly bodies, in contrast to the movement of terrestrial elements in *De Caelo et Mundo*.

3. P. Virgilius Maro, *Aeneid*, 4.201–2 (*ipse diem noctemque negat discernere caello / nec meminisse viae media Pallinurus in unda*), 4.585–87 (*nam neque erant astrorum ignes nec lucidus aethra / siderea polus, obscuro sed nubila caelo, / et lunam in nimbo nox intempesta tenebat*).

4. Peter Bullions, *A Copious and Critical Latin-English Dictionary: Abridged and Re-arranged* (New York: Sheldon, 1866), p. 183.

5. Federal Aviation Administration (FAA), "Busting Myths about the FAA and Unmanned Aircraft," March 7, 2014.

6. Gregory S. McNeal, "The Federal Government Thinks Your Backyard Is National Airspace and Toys Are Subject to FAA Regulations," *Forbes*, November 18, 2014.

7. Administrator's Response to Respondent's Motion to Dismiss at 4–5, *Huerta v. Pirker*, No. CP-217 (NTSB Nov. 1, 2013), at 5, http://www.suasnews.com/wp-content/uploads/2013/11/FAA_Response.pdf.

8. C. Lincoln Bouvé, "Private Ownership of Airspace," *Air Law Review* 1 (1930): 232; Smith, "The Origins of Trespass."

9. *Bury v. Pope*, (1587) 1 Cro Eliz 118, 78 ER 375, n.8 ("*Nota: cuius est solum ejus est summitas usque ad coelum*. Temp Ed 1.").

10. Edward Coke, "Of Real Property, and First or Corporeal Hereditaments of Land," *The First Part of the Institutes of the Lawes of England* (London: Rawlins, Roycroft, and Sawbridge, 1684).

11. 22 Henry VI 59; 10 Edward IV 14.

12. 14 Henry VIII 12.

13. George B. Logan, "Aviation and the Maxim *Cujus Est Solum*," *St. Louis Law Review* 16 (1931): 303, 306. Note that much of the text in the article was taken from *Swetland v. Curtiss Airports Corporation*, 41 F.2d 929 (N.D. Ohio, 1930).

14. Logan, "Aviation and the Maxim *Cujus Est Solum*," p. 305.

15. Penruddock's Case, Trin 40 Eliz 101 [3 Coke 205 (1597)].

16. Reports of Sir Edward Coke, Knt: in English, in Thirteen Parts, vol. 3, Penruddock's Case, Trin 40 Eliz 101.

17. Baten's Case, 9 Coke 54 (1611). See also *Bury v. Pope*, (1587) Cro Eliz 118 [1653], ER 382, (1653) Cro Eliz 118, (1653) 78 ER 375 (B) (citing the maxim).

18. William Blackstone, *Commentaries on the Laws of England*, vol. 2 (London, G.W. Childs, 1866), p. 18.

19. William Blackstone, *Commentaries on the Laws of England*, vol. 2 (Philadelphia, J.B. Lippincott & Co., 1893), p. 18.

20. Blackstone, *Commentaries*, vol. 2 (1893), p. 18.

21. *Pickering v. Rudd*, (1815) 4 Camp 219 (writing, "Nay, if this board overhanging the plaintiff's garden be a trespass, it would follow that the aeronaut is liable to an action of trespass *quare clausum fregit* at the suit of the occupier of every field over which his balloon passes in the course of his voyage.").

22. *Fay v. Prentice and Another*, [1845] ER 79, (1845) 1 CB 828, (1845) 135 ER 769.

23. 1 CB 838 (Coltman, J.).

24. 1 CB 838–41 (Maule, J.); 1 CB 840–41 (Cresswell, J.).

25. *Electric Telegraph Co. v. Overseers of the Poor of the Township of Salford*, 11 Ex 181 (1855).

26. 11 Ex 189.

27. *Kenyon v. Hart*, [1865] EWHC QB J102; 122 ER 1188; 6 B & S 249; 32 LJR 87 (1865).

28. EWHC QB J102.

29. *Corbett v. Hill*, (1870) LR 9 Eq 671, (1870) 39 CJCh 547, (1870) 2 LT 263, (1870) 7 Digest (Repl) 267.

30. LR 9 Eq 671.

31. *Ellis v. Loftus*, (1874) LR 10, CP 10.

32. *Clifton v. Viscount Bury and Others*, (1887–1888) 4 TLR 1.

33. 4 TLR 1.

34. 4 TLR 1.

35. 4 TLR 1, citing *Pickering v. Rudd*, 1, Stark, 58. See also *Kenyon v. Hart*, 6 B. and S. 249.

36. 4 TLR 1.

37. *Lemmon v. Webb*, [1895] AC 1, Bailii, [1894] UKHL 1. See also *Lemmon v. Webb*, [1894] 3 Ch 1.

38. Frederick Pollock, *The Law of Torts*, 1st ed. (London: Stevens and Sons, 1887), *280, p. 218.

39. *Entick v. Carrington*, (1765) 19 St Tr 1066, cited in Pollock, *Law of Torts*.

40. Pollock, *Law of Torts*, 1st ed., p. 218.

41. Pollock, *Law of Torts*, 1st ed., p. 218 (footnotes omitted).

42. Pollock, *Law of Torts*, 1st ed., p. 219.

43. Pollock, *Law of Torts*, 1st ed., p. 219.

44. Pollock, *Law of Torts*, 1st ed., p. 219.

45. Pollock, *Law of Torts*, 1st ed., p. 219 (emphasis added).

46. Pollock, *Law of Torts*, 1st ed., p. 219.

47. Pollock, *The Law of Torts*, 11th ed. (London: Stevens and Sons, 1920), p. 350.

48. Pollock, *Law of Torts*, 11th ed., p. 350.

49. Pollock, *Law of Torts*, 11th ed., p. 350–352.

50. *Finchley Electric Light Co. v. Finchley Urban Dist.*, [1903] 1 Ch 437 (overruled [1902] 1 Ch 866).

51. 1 Ch 437.

52. James Kent, *Commentaries on American Law*, vol. 1 (St. Paul, Minnesota: West Publishing Co., 1894), p. 188. See also *State v. Campbell, T.U.P. Charlton (Ga.) 166* ("When the American Colonies were first settled by our ancestors it was held, as well by the settlers, as by the judges and lawyers of England, that they brought hither as a birthright and inheritance so much of the common law as was applicable to their local situation, and change of circumstances."); (1722) 2 Peere Wms 75 ("9th August 1722, it was said by the Master of the Rolls to have been determined by the Lords of the Privy Council upon an appeal to the King in council from the foreign plantations, 1st, that if there be a new and uninhabited country found out

by English subjects, as the law is the birthright of every subject, so wherever they go they carry their laws with them, and, therefore, such new-found country is to be governed by the laws of England.").

53. Richard C. Dale, "The Adoption of the Common Law by the American Colonies," *American Law Register* 30, no. 9 (1882): 553, 554.

54. Massachusetts Constitution (1780), ch. VI, art. VI ("All the laws which have heretofore been adopted, used and approved in the Province, Colony or State of Massachusetts Bay, and usually practiced on in the Courts of law, shall still remain and be in full force, until altered or repealed by the Legislature; such parts only excepted as are repugnant to the rights and liberties contained in this Constitution.").

55. Code of Virginia 1860, ch. 16, § 1 ("The common law of England, so far as it is not repugnant to the principles of the Constitution of this state shall continue in force within the same.").

56. See Vermont, ch. 32, General Statutes of 1870. See also North Carolina Code 1855, ch. 4, § 1. ("All such parts of the common law as were heretofore in force and use within this state, or so much of the common law as is not destructive of or repugnant to, or inconsistent with, the freedom and independence of this state and the form of government therein established, and which has not been otherwise provided for in the whole or in part, not abrogated, repealed or become obsolete, are hereby declared to be in full force within the state.").

57. California Act of April 13, 1850, Gen. Laws, p. 599.

58. See Illinois Rev. Stat. 1874, ch. 28, § 1 ("That the common law of England, so far as the same is applicable and of a general nature, and all statutes or acts of the British Parliament made in aid of or to supply the defects of the common law prior to the fourth year of James I [with some exceptions] and which are of a general nature and not local to that kingdom, shall be the rule of decision, and shall be considered as of full force until repealed by legislative authority."); Indiana Act of May 31, 1852 (same wording as the Illinois statute); Kansas Rev. Stat. 1868, ch. 119, § 3 ("The common law, as modified by constitutional and statutory law, judicial decisions and the condition and wants of the people shall remain in force in aid of the general statutes of the state."); Missouri Rev. Stat. 1870, ch. 86, § 1 ("The common law of England and all statutes and Acts of Parliament made prior to the fourth year of the reign of James I, and which are of a general nature and not local to that kingdom, which common law and statutes are not repugnant to or inconsistent with the Constitution of the United States, the Constitution of this state, or the statute laws in force for the time being, shall be the rule of action and decision in this state, any law, custom or usage to the contrary notwithstanding."); Nebraska Rev. Stat. 1873, § 1 ("So much of the common law of England as is applicable and not inconsistent with the Constitution of the United States, the constitution of this state or with any law passed or to be passed by the legislature

thereof is adopted and declared to be the law within this state."); Wisconsin Constitution, § 13 ("Such parts of the common law as are now in force in the territory of Wisconsin not inconsistent with this Constitution shall be and continue part of the law of this state until altered or suspended by the legislature.").

59. *Lyman v. Bennett*, 8 Mich. 18.

60. 8 Mich. 18.

61. See, for example, *Grandone v. Lovdal*, 70 Cal. 161 (1866); *Tanner v. Wallbrun*, 77 Mo. App. 262 (1898); *Countryman v. Lighthill*, 24 Hun. (N.Y.) 405 (1881).

62. See, for example, *Aiken v. Benedict*, 39 Barb. (N.Y.) 409 (1863); *Smith v. Smith*, 110 Mass. 302 (1872); *Harrington v. McCarthy*, 169 Mass. 492 (1897); *Lawrence v. Hough*, 35 N.J. Eq. 371 (1882).

63. See, for example, Reimer's Appeal, 100 Pa. Sta. 182 (1882) (bay window); *Murphy v. Bolger*, 60 Vt. 723 (1888) (roof); *Wilmarth v. Woodcock*, 58 Mich. 482, 25 N.W. 475 (1885); *Lawrence v. Hough*, 35 N.J. Eq. 371 (1882) (cornices); *Langfeldt v. McGrath*, 33 Ill. App. 158 (1889); *Codman v. Evans*, 7 Allen (Mass.) 431 (1863) (wall); *Lyle v. Little*, 83 Hun. 532, 33 N.Y.S. 8 (1895) (wall); *Smith v. Smith*, 110 Mass. 302 (1872); *Esty v. Baker*, 48 Me. 495 (1860); *Huber v. Stark*, 124 Wis. 359, 102 N. W. 12 (recognizing both trespass and nuisance); *Corner v. Woodfill*, 126 Ind. 85, 25 N. E. 876 (1890). *Cf. Zander v. Valentine Blatz Brewing Co.*, 95 Wis. 162, 70 N. W. 164 (1879); *Rahn v. Milwaukee Electric Railway & Light Co.*, 103 Wis. 467, 79 N. W. 747 (1899).

64. See, for example, *Meyer v. Metzler*, 51 Cal. 142 (1875); *Wilmarth v. Woodcock*, 58 Mich. 482, 25 N. W. 475 (1885); *Norwalk Heating & Lighting Co. v. Vernam*, 75 Conn. 662, 55 Atl. 168 (1903); *Barnes v. Berendes*, 139 Cal. 32, 69 Pac. 491 (1902); *Copper v. Dolvin*, 68 Ia. 757 (1886); *Langfeldt v. McGrath*, 33 Ill. App. 158 (1889); *Kafka v. Bozio*, 191 Cal. 746, 218 Pac. 753 (1923).

65. Edward C. Sweeney, "Adjusting the Conflicting Interests of Landowner and Aviator in Anglo-American Law," *Journal of Air Law and Commerce* 3 (1932): 534–35, n.150 (comparing cases where an action of ejectment was permitted as against those where it was denied).

66. *Metropolitan West Side Elevated R.R. Co. v. Springer*, 171 Ill. 170, 175, 49 N.E. 416 (1897).

67. 171 Ill. 175.

68. See *Atkins v. Bordman*, 2 Met. (43 Mass.) 457, 467 (1841) (noting that according to *Cujus est solum, ejus est usque ad coelum*, the owner of an estate may make any and all beneficial uses of it as their own pleasure and may alter the mode of using it by erecting or removing building over it or digging into or under it without restraint).

69. See *Baldwin v. Breed*, 16 Conn. 60, 66 (1843) (applying *Cujus est solum, ejus est usque ad coelum*, by stating that separating a building from the ground on which it stands cannot be carried into effect without great difficulty); *Becket v. Clark*, 40

Conn. 485, 488 (1873) (stating that *Cujus est solum, ejus est usque ad coelum* makes it so that the house and land are understood to be divided based upon the same line); *Isham v. Morgan*, 9 Conn. 374, 377 (1832) (Consistent with *cujus est solum, ejus est usque ad coelum*, the delivery of the deed to Lydia conveyed not only the land but the buildings upon the land.); *Stevenson v. Bachrach*, 170 Ill. 253, 48 N. E. 327, 328 (1897) (holding that the building which rests upon the portion of the land owned by appellee belongs in severalty to him, and the portion of the building which rests upon the land belonging to appellants belongs in severalty to them).

70. See *Barnett v. Johnson*, 15 N.J. Eq. 481, 489 (1856-abbreviated) (Defendant proposed to build an arch and a building over the canal that would block the windows of the plaintiff who owned the adjacent property, thus restricting light and air to the property; the court ruled that the plaintiff had the right to light and air.). But see *Mahan v. Brown*, 13 Wend. (N.Y.) 261, 263 (1835) (right to light waived by building too close to the property line).

71. See *Gas Products Co. v. Rankin*, 63 Mont. 372, 389, 207 Pac. 993, 997, 24 A. L. R. 294 (1922-in quotation from Blackstone) (finding chapter 125 of the Laws of 1921, prohibiting "wasteful and extravagant" uses of natural gases, to be unconstitutional as petroleum and gas beneath the surface belong to the landowner); *Chase v. Silverstone*, 62 Me. 175, 183 (1873-abbreviated) (protecting a landowner's right to dig a well); *Hague v. Wheeler*, 157 Pa. St. 324, 27 At. 714 (1892-Eng. equiv.) (The mere fact that the defendants, by operations upon their land, are taking gas from the earth, and thereby diminishing the quantity of gas which would otherwise come to the plaintiffs' wells, furnishes no ground for complaint or equitable interference); *Lenfers v. Henke*, 73 Ill. 405, 408 (1874-Eng. equiv.) (property ownership conveying right to lead mines); *Lime Rock R. R. Co. v. Farnsworth*, 86 Me. 127, 29 At. 957, 958 (1893-Eng. equiv.) (property ownership conveying marble and lime rock within the tract); *Louisville & Nashville R. R. Co. v. Boykin*, 76 Ala. 560, 563 (1884-Eng. equiv.) (conveying right to gravel along with purchase of land); *Pence v. Carney*, 58 W.Va. 296, 52 S. E. 702, 6 L.R.A. (n. s.) 266 (1905) (right to percolating water). But see *Katz v. Walkinshaw*, 141 Cal. 116, 74 Pac. 766, 64 L.R.A. 236 (1902) (applying the doctrine of reasonable use as a limit on *ad coelum* for well water "to such amount of water as may be necessary for some useful purpose in connection with the land from which it is taken"); *Meeker v. East Orange*, 77 N.J.L. 623, 636, 74 At. 379, 384, 25 L.R.A. (n. s.) 465 (1909) (limiting sale of percolating water).

72. See *First Baptist Society v. Wetherell*, 34 R.I. 155, 82 Atl. 1061, 1063 (1912).

73. See, for example, *Huber v. Stark*, 124 Wis. 359 (1905) (eaves entering adjacent airspace held to be a trespass); *Crocker v. Manhattan Life Insurance Co.*, 61 App. Div. (N.Y.), 226 (1901) (eaves extending over an adjacent owner's land held to be a trespass); *Ackerman v. Ellis*, 81 N.J.L. 1 (1911) (holding branches extending into a property owner's airspace to be a nuisance); *Puorto v. Chieppa*, 78 Conn. 401

(1905) (holding that a board attached to building extending to adjacent airspace constituted a trespass); *Norwalk Heating Lighting Co. v. Vernam*, 75 Conn. 662 (1903) (a wooden structure extending over an adjacent property amounted to a trespass); *Whittaker v. Standvick*, 100 Minn. 396 (1907) (shooting over a pass between two navigable lakes ruled both a trespass and a nuisance); *Crocker v. Manhattan Life Insurance Co.*, 61 App. Div. 226, 70 N.Y.S. 492 (1901) (cornice projecting into adjacent property considered a trespass); *Barnes v. Berendes*, 139 Cal. 32, 69 P. 491, 72 P. 406 (1902) (wall extending into neighbor's property considered a trespass); *Barnes v. Berendes*, 139 Cal. 32 (1903) (leaning walls considered a trespass); *Herrin v. Sutherland*, 241 P. 328 (Mont. 1925) (shooting of wild ducks above a navigable stream); *Portsmouth Co. v. United States*, 260 U.S. 327 (1922) (on appeal from the Court of Claims) (firing of guns over private property); *Milton v. Puffers*, 207 Mass. 416, 93 N. E. 634 (1911) (protruding foundation stones considered both a trespass and a nuisance).

74. *Young v. Thiedick*, 28 Oh. Ct. Ap. 239 (1918).

75. *Herrin v. Sutherland*, 74 Mont. 587, 241 Pac. 328 (1925) (confirming that airspace over one's property is subject to protection against trespass).

76. 204 Pac. 328.

77. *Hannaballson v. Session*, 116 Iowa, 457 (1902).

78. See, for example, *Winton v. Cornish*, 5 Ohio 478 (1931); *Frazier v. Brown*, 12 Ohio Stat. 294, 304 (1891); *Winslow v. Fuhrman*, 25 Ohio Stat. 639, 651 (1904).

79. *Portsmouth Harbor Land & Hotel Co. v. United States*, 260 U.S. 327 (1922).

80. 260 U.S. 329–30. For two prior cases finding insufficient sustained engagement to constitute a taking, see *Peabody v. United States*, 231 U.S. 530 (1913), and *Portsmouth Harbor Land & Hotel Co. v. United States*, 250 U.S. 1 (1919).

81. *Butler v. Frontier Telephone Co.*, 186 N.Y. 486, 491 (1906) (Vann, J.).

82. 186 N.Y. 492.

83. *Penn. Coal Co. v. Mahon*, 260 U.S. 393, 419 (1878).

84. 260 U.S. 393, 419.

85. See *Smith v. City of Atlanta*, 92 Ga. 119, 17 S. E. 981 (1893).

86. Eugene McQuillin, *A Treatise on the Law of Municipal Corporations*, vol. 8 (Chicago: Callaghan and Co., 1921), § 2775 (internal quotations omitted). See also McQuillin, *Municipal Corporations*, vol. 6 (1913), § 2775; *Incorporated Town v. Cent. States El. Co.*, 204 Iowa 1246, 1250, 214 N.W. 879, 54 A.L.R. 474 (1927) (quoting McQuillin).

87. Byron Elliott and William Elliott, *A Treatise on the Law of Roads and Streets*, vol. 2, 3rd ed. (Indianapolis, IN: Bobbs-Merrill Co., 1911), p. 256, § 830.

88. Elliott and Elliott, *Law of Roads and Streets*, p. 257.

89. *Wheeler v. City of Ft. Dodge*, 131 Iowa 566 (1906) ("The fact that the wire in most of its course passed through the air above the heads of the people using the

walks and carriageway below does not remove its character as an obstruction of the street. The public right goes to the full width of the street, and extends indefinitely upward and downward, so far at least as to prohibit encroachment upon said limits by any person by any means by which the enjoyment of said public right is or may be in any manner hindered or obstructed or made inconvenient or dangerous.")

90. See, for example, *Bohen v. Waseca*, 32 Minn. 176 (19 N.W. 730, 50 Am. Rep. 564) (1884) (awning); *Hume v. Mayor*, 74 N.Y. 264 (1878), and *Drake v. Lowell*, 13 Met. (Mass.) 292 (1847) (awning); *Bybee v. State*, 94 Ind. 443 (48 Am. Rep. 175) (1905) (bridge or covered viaduct); *Jones v. City of New Haven*, 34 Conn. 1 (1867) (tree); Reimer's Appeal, 100 Pa. 182 (45 Am. Rep. 373) (1882) (bay window).

91. See, for example, *Beall v. Seattle*, 28 Wash. 593 (69 P. 12, 61 L.R.A. 583, 92 Am. St. Rep. 892) (1902) (boiler underground). See also *Abilene v. Cowperthwait*, 52 Kan. 324 (34 P. 795) (1903); *Smith v. Leavenworth*, 15 Kan. 81 (1875); *Woodbury v. District of Columbia*, 5 Mackey (D.C.) 127 (1886).

92. *Incorporated Town v. Cent. States El. Co.*, 214 N.W. 879 (Iowa 1927).

93. Articles of Confederation, art. II.

94. Federalist No. 45 (James Madison), 1788, reprinted in Alexander Hamilton, James Madison, John Jay, Clinton Rossiter, *The Federalist Papers* (New York: Penguin, 1961), p. 291.

95. Federalist No. 16 (Alexander Hamilton), 1787, reprinted in Alexander Hamilton, James Madison, John Jay, Clinton Rossiter, *The Federalist Papers* (New York: Penguin, 1961), p. 115.

96. Federalist No. 45 (Madison), pp. 292–93.

97. U.S. Const. art. I, § 8, cl. 17. See also Federalist No. 43 (James Madison), 1788, reprinted in Alexander Hamilton, James Madison, John Jay, Clinton Rossiter, *The Federalist Papers* (New York: Penguin, 1961), pp. 272–73 ([The federal district is] to be appropriated to this use with the consent of the State ceding it; as the State will no doubt provide in the compact for the rights and the consent of the citizens inhabiting it; . . . as they will have had their voice in the election of the government which is to exercise authority over them; . . . as the authority of the legislature of the State, and of the inhabitants of the ceded part of it, to concur in the cession will be derived from the whole people of the State in their adoption of the Constitution, every imaginable objection seems to be obviated.").

98. U.S. Const. art. IV, § 3.

99. U.S. Const. art. IV, § 4.

100. U.S. Const. art. IV, § 2.

101. U.S. Const. amend. X.

102. *Jacobson v. Massachusetts*, 197 U.S. 11, 25 (1905) (internal quotations omitted).

103. *Thomas Cusack Co. v. City of Chicago*, 242 U.S. 526, 531 (1917).

104. *United States v. Bevans*, 16 U.S. 336, 389 (1818).

105. *Commonwealth v. Young*, Brightly N.P., 302, 309 (Pa. 1818).

106. *New York v. Miln*, 36 U.S. 102, 139 (1837).

107. Act for the Admission of Alabama, March 2, 1819, 15th Cong., Sess. II, ch. 47, pp. 489–492, § 6, 3.

108. *Pollard v. Hagan*, 44 U.S. 212, 221 (1845).

109. 44 U.S. 222.

110. U.S. Const. art. I, § 8, cl. 3.

111. *Gibbons v. Ogden*, 22 U.S. (9 Wheat.) 1, 195 (1824).

112. 22 U.S. at 195.

113. 22 U.S. at 203.

114. 22 U.S. at 203.

115. Submerged Lands Act of 1953, Pub. L. 31, May 22, 1953, codified at 43 U.S.C. §§ 1301–1315.

116. See, for example, *United States v. Pennsylvania Salt Manufacturing Co.*, 16 F.2d 476, 479 (E.D. Pa. 1926) (1926); Appeal of York Haven Water & Power Co., 212 Pa. 622, 62 A.97, 97 (1905); *Wainwright v. McCullough*, 66 (1869*); Commonwealth* ex rel. *Hensel v. Young Men's Christian Association of Warren*, 169 Pa. St. 24, 121 (1895).

117. See *Cochran v. Preston*, 108 Md. 220, 70 Atd. 113 (1908-Eng. equiv.).

118. *See Horan v. Byrnes*, 72 N.H. 93, 54 Atl. 945, 62 L.R.A. 602 (1903).

119. See, for example, *Village of Euclid v. Ambler Realty Co.*, 272 U.S. 365, 395 (1926) (recognizing that zoning ordinances can only be declared unconstitutional where they are "clearly arbitrary and unreasonable, having no substantial relation to the public health, safety, morals, or general welfare").

120. See, for example, *Greene v. Town of Blooming Grove*, 879 F.2d 1061, 1063 (1989), citing *City of Cleburne v. Cleburne Living Center*, 473 U.S. 432, 440 (1985).

121. See *Thomas Cusack Co. v. Chicago*, 242 U.S. 526 (1917).

122. 242 U.S. at 530–31.

123. Earl N. Findley, "Twenty Months of Commercial Aeronautics," *New York Times*, January 16, 1921.

124. Bogert, *Problems in Aviation Law*, pp. 271–72.

125. Bogert, *Problems in Aviation Law*, pp. 273–73.

126. See, for example, William R. McCracken, "Air Law," *American Law Review* 57 (1923): 99; Edmund F. Trabue, "The Law of Aviation," *American Law Review* 58 (1924): 65; "Current Legislation: The Air Commerce Act of 1926," *Columbia Law Review* 27, no. 8 (1927): 989.

127. *Guille v. Swan*, 19 Johns. 381 (N.Y.) (1822).

128. *Bank of Monongahela Valley v. Weston,* 159 N.Y. 201, 54 N.E. 40, 45 L.R.A. 547 (1912), in 19 Case and Comment 681 (1913), 42 A.L.R. 951.

129. U.S. Const., art. III, § 2(1). For accounts of different constitutional theories for federal regulation of air travel, see Henry Hotchkiss, *A Treatise on Aviation*

Law (New York: Baker, Voorhis and Co., 1928), §§ 54–57; Carl Zollman, *Law of the Air* (Milwaukee, WI: Bruce Publishing Co., 1927), § 54 et seq.

130. See, for example, The Crawford Bros. No. 2, 215 Fed. 269 (W.D. Wash. 1914) (jurisdiction declined for repairs against an airplane which had fallen into navigable waters under federal maritime jurisdiction); The Steamer St. Lawrence, 1 U.S. 522, 527 (1861) ("[N]o State law can enlarge [admiralty jurisdiction] nor can an act of Congress or rule of court make it broader than the judicial power may determine to be its true limits.").

131. Zollman, *Law of the Air*, 39 ff; "Air Commerce Act," *Columbia Law Review*: 989, n.4.

132. Zollman, *Law of the Air*, 39 ff; "Air Commerce Act," *Columbia Law Review*: 989, n.4.

133. American Bar Association, Report of the Forty-Fourth Annual Meeting of the American Bar Association held at Cincinnati, Ohio, Aug. 31, Sept. 1 and 2, 1921 (Cincinnati, Ohio: Lord Baltimore Press, 1921): 81 ("the best ultimate solution of this question is one that will put the United States on a par with other nations, and that is a constitutional amendment, which will extend the power of Congress to legislate on flight through the air.").

134. Rowan A. Greer, "International Aerial Regulations," Air Service Information Circular (Washington: Chief of the Air Service, 1926), p. 29.

135. William A. Schnader, "Uniform Aviation Liability Act," *Journal of Air Law and Commerce* 9 (1938): 664. For discussion of the evolution of the model statute, see George Gleason Bogert, "Recent Developments in the Law of Aeronautics," *Cornell Law Quarterly* 8 (1923): 26. See also Herzel H. E. Plaine, "State Aviation Legislation," *Journal of Air Law and Commerce* 14 (1947): 333, 334 (referring to the Uniform Aeronautics Act of 1922 as "[p]erhaps the most influential of the uniform acts.").

136. Uniform State Law for Aeronautics, National Conference of Commissioners on Uniform State Laws, conference, San Francisco, August 1922, 11 Uniform Laws Anno. 159, § 3 ("The ownership of the space above the lands and waters of this state is declared to be vested in the several owners of the surface beneath, subject to the right of flight described in Sec. 4."). Hereinafter "Uniform Aeronautics Act of 1922."

137. Uniform Aeronautics Act of 1922, § 3. See also § 4 ("Flight in aircraft over the lands and waters of this state is lawful, unless at such low altitudes as to interfere with the then existing use to which the land or water, or the space over the land or water, is put by the owner, or unless so conducted as to be imminently dangerous to persons or property lawfully on land or water beneath.").

138. Uniform Aeronautics Act of 1922, 11 Uniform Laws Anno. 159.

139. Uniform Aeronautics Act of 1922, §§ 4, 5.

140. Uniform Aeronautics Act of 1922, § 5.

141. Schnader, "Uniform Aviation Liability Act."

142. Mass. Gen. Laws ch. 534 (1922) ("An Act Regulating the Operation of Aircraft").

143. Mass. Gen. Laws ch. 534.

144. *Smith v. New England Aircraft Co.* [Mass.] 170 N.E. 385 (1930). For a scathing critique of the court's opinion in this case, see George B. Logan, "The Case of Smith v. New England Aircraft Company," *Journal of Land and Public Utility Economics* 6 (1930): 316–24.

145. 170 N.E. 385.

146. 170 N.E. 385.

147. 170 N.E. 385.

148. 170 N.E. 385.

149. 170 N.E. 385.

150. 170 N.E. 385.

151. Civil Aviation, H.R. Rep. No. 69-572, at 7 (1926).

152. Air Commerce Act, Pub. L. No. 69-254, 44 Stat. 568 (1926). For discussion of the law prior to enactment of the statute, see Frederick P. Lee, "Air Commerce Act of 1926," *American Bar Association Journal* 12 (1926): 371; Roger F. William, "Federal Legislation Concerning Civil Aeronautics," *University of Pennsylvania Law Review* 76 (1928): 798.

153. H.R. Rep. No. 69-572, at 7.

154. H.R. Rep. No. 69-572, at 8.

155. 44 Stat. 568, 569–70.

156. 44 Stat. 572-573, § 7.

157. 44 Stat. 571, § 5(b).

158. 44 Stat. 574 § 10. See also 44 Stat. 570, § 3(e), codified at 49 U.S.C.A. §173(e) ("Regulatory Powers. The Secretary of Commerce shall by regulation . . . (e) Establish air traffic rules for the navigation, protection, and identification of aircraft, including rules as to safe altitudes of flight and rules for the prevention of collisions between vessels and aircraft.").

159. H.R. Rep. No. 69-572, at 9 ("[A]ir space, with its absence of fixed roads and tracks and aircraft with their ease of maneuver, present as to transportation practical and legal problems similar to those presented by transportation by vessels upon the high seas.").

160. H.R. Rep. No. 69-572, at 10.

161. In 1910 Congress created the Bureau of Lighthouses in the Department of Commerce. 36 Stat. 537 (1910). On July 1, 1939, the president's Reorganization Plan No. 2 transferred the agency to the Coast Guard. 4 F.R. 2731, 53 Stat. 1431. The U.S. Coast and Geodetic Survey was first established in 1878 as part of the Department of the Treasury. Act of June 20, 1878 (20 Stat. 215). In 1903, it

was transferred to the Department of Commerce and Labor by the act creating that department, and a decade later to the Department of Commerce. 32 Stat. 825 (1903); 37 Stat. 736 (1913). In 1965, it became folded into the Environmental Science Services Administration (ESSA), Department of Commerce. Reorganization Plan No. 2 of 1965, effective July 13, 1965. The organization was abolished by Reorganization Plan No. 4 of 1970, effective October 3, 1970, which transferred personnel and functions of ESSA to the newly established National Oceanic and Atmospheric Administration, Department of Commerce.

162. H.R. Rep. No. 69-572, at 9.

163. H.R. Rep. No. 69-572, at 9 ("The declaration of what constitutes navigable air space is an exercise of the same source of power, the interstate commerce clause, as that under which Congress has long declared in many acts what constitutes navigable or nonnavigable waters.").

164. H.R. Rep. No. 69-572, at 9 ("The public right of flight in the navigable air space owes its source to the same constitutional basis which, under decisions of the Supreme Court, has given rise to a public easement of navigation in the navigable waters of the United States, regardless of the ownership of the adjacent or subjacent soil.").

165. Convention Relating to the Regulation of Aerial Navigation, Oct. 13, 1919, art. I. ("The High contracting Parties recognize that every Power has complete and exclusive sovereignty over the air space above its territory."). Hereinafter "1919 International Convention." See also H.R. Rep. No. 69-572, at 10 (citing the 1919 convention).

166. 1944 Chicago Convention on International Civil Aviation, ratified by the United States February 20, 1946.

167. 1919 International Convention, ch. 1.

168. Inter-American Commercial Aviation Convention (February 20, 1928), art. I. ("The high contracting parties recognize that every state has complete and exclusive sovereignty over the air space above its territory and territorial waters."). Senate advice and consent to ratification February 20, 1931; ratified by the president March 6, 1931; terminated as to the United States November 29, 1947, 47 Stat. 1901.

169. Inter-American Commercial Aviation Convention, art. 5.

170. Inter-American Commercial Aviation Convention, arts. 7, 9.

171. 1929 Protocol to the International Convention Relating to the Regulation of Aerial Navigation, June 15, 1929 (Paris), art. 3.

172. 1929 Protocol, art. 3.

173. 1929 Protocol, ch. IV.

174. Air Commerce Regulations (December 1926), Chapter 7, § 74(G).

175. Regulation of Air Craft, Session Laws of Wyoming, 1927, ch. 72, sec. 2, j.

176. General Act Relative to Aeronautics, June 26, 1929, 113 Ohio Laws, p. 28.

177. Early 20th century Ohio history is interwoven with the Wright brothers and the growth of modern aviation. Orville Wright was born in Dayton in 1871, where both he and his older brother Wilbur attended school. In 1889 they became local business owners, opening a print shop and publishing a local newspaper using a printing press that they designed and built. A few years later, they opened a bike shop and began manufacturing bicycles. The brothers went on to test planes at Huffman Field near Dayton and in 1909 founded the Wright Company there. Although Wilbur died, Orville carried on and became a leading national figure, serving, inter alia, as an adviser to the National Advisory Committee for Aeronautics—the predecessor to the National Aeronautics and Space Administration. See generally "Wright Brothers," Ohio History Central, Ohio History Connection. Because of the Wright brothers, "Ohio regards itself as the mother state of aviation." *Swetland v. Curtiss Airports Corp.,* 41 F.2d 929, 932 (N.D. Ohio 1930). Ohio was among the first states to pass laws to provide for municipal airports. See Ohio General Code, § 3677; State ex rel. *Chandler v. Jackson,* 121 Ohio St. 186, 167 N.E. 396 (1929). And the issuance of bonds for that purpose. Ohio General Code, § 3939; *State v. City of Cleveland,* 26 Ohio App. 265, 160 N.E. 241 (1937).

178. 41 F.2d 929.

179. 41 F.2d at 932.

180. 41 F.2d at 932.

181. 41 F.2d at 932.

182. 41 F.2d at 933.

183. 41 F.2d at 933.

184. 41 F.2d at 933.

185. 41 F.2d at 935.

186. 41 F.2d at 935.

187. 41 F.2d at 937, citing and quoting Hiram L. Jome, "Property in the Air as Affected by the Airplane and the Radio," *American Law Review* 62 (1928): 894–95.

188. 41 F.2d at 937.

189. 41 F.2d at 937.

190. 41 F.2d at 938.

191. 41 F.2d at 943.

192. *Swetland v. Curtiss Airports Corporation,* 55 F.2d 201, 203 (6th Cir. 1932).

193. 55 F.2d at 203.

194. 55 F.2d at 203.

195. 55 F.2d at 203.

196. 55 F.2d at 204–5.

197. 55 F.2d at 205.

198. Uniform Licensing Act of 1930, 11 Uniform Laws Anno. 185.

199. Uniform Aeronautical Regulatory Act, 11 Uniform Laws Anno. 173.

200. 11 Uniform Laws Anno. 193.

201. Plaine, "State Aviation Legislation," p. 335.

202. Logan, "Aviation and the Maxim *Cujus Est Solum*," p. 304.

203. See, for example, Arnold D. McNair, *The Law of the Air* (London: Butterworth and Co., 1932); Logan, "Aviation and the Maxim *Cujus Est Solum*," p. 306; Arthur L. Newman, II, "Aviation Law and the Constitution," *Yale Law Journal* 39 (1930): 1113; Carl Zollmann, "Aircraft as Common Carriers," *Journal of Air Law and Commerce* 1, no. 2 (1930): 190; Reed G. Landis, "State Agencies of Control and Enforcement of Aeronautical Laws," *Journal of Air Law and Commerce* 1, no. 2 (1930): 186; Fred D. Fagg Jr., "Incorporating Federal Law into State Legislation," *Journal of Air Law and Commerce* 1, no. 2 (1930): 199; Chester W. Cuthel, "Development of Aviation Laws in the United States," *Air Law Review* 1 (1930): 86; Sweeney, "Adjusting the Conflicting Interests of Landowner and Aviator in Anglo-American Law," pp. 531–616.

204. Civil Aeronautics Act of 1938, S. Rep. No. 75-1661, at 2 (1938).

205. S. Rep. No. 75-1661, at 2.

206. S. Rep. No. 75-1661, at 2.

207. S. Res. 146, 74th Cong., 1st sess. (1935).

208. S. Rep. No. 75-1661, at 2 (Senator Royal Copeland)

209. Civil Aeronautics Act of 1938, § 1107(i)(3).

210. See Air Commerce Act of 1926, Pub. L. 69-254, § 6 (stating "that the Government of the United States has, to the exclusion of all foreign nations, complete sovereignty of the airspace over the lands and waters of the United States.").

211. *Braniff Airways, Inc. v. Nebraska State Board of Equalization & Assessment,* 347 U.S. 590, 595 (1954). See also H.R. Rep. No. 69-572, at 10.

212. See Uniform Aeronautics Act § 4 (Uniform Law Commission 1922), § 2 ("Sovereignty in the space above the lands and waters of this State is declared to rest in the State, except where granted to and assumed by the United States pursuant to a constitutional grant from the people of this State."), § 3 ("The ownership of the space above the lands and waters of this State is declared to be vested in the several owners of the surface beneath, subject to the right of flight described in Section 4."), § 4 ("Flight in aircraft over the lands and waters of this State is lawful, unless at such a low altitude as to interfere with the then existing use to which the land or water, or the space over the land or water, is put by the owner, or unless so conducted as to be imminently dangerous to persons or property lawfully on the land or water beneath. The landing of an aircraft on the lands or waters of another, without his consent, is unlawful, except in the case of a forced landing. For damages caused by a forced landing, however, the owner or lessee of the aircraft or the aeronaut shall be liable. . . .").

213. Civil Aeronautics Act of 1938, 52 Stat. 973, §§ 601–10.

214. 52 Stat. 973, §§ 701–2 (air safety board); §§ 901–3 (civil penalties, criminal penalties, and venue for prosecution).

215. 52 Stat. 973, §§ 401–16.

216. 52 Stat. 973, §§ 201–6.

217. First Annual Report of the Civil Aeronautics Authority: Fiscal Year Ended June 30, 1939, with Additional Activities to November 1939 (Washington: Government Printing Office, 1940), pp. 39–41.

218. Second Annual Report of the Civil Aeronautics Authority: Fiscal Year Ended June 30, 1940 (Washington: Government Printing Office, 1940), pp. 1–14.

219. Second Annual Report of the Civil Aeronautics Authority, pp. 14–22.

220. Second Annual Report of the Civil Aeronautics Authority, pp. 22–23.

221. Second Annual Report of the Civil Aeronautics Authority, p. 5.

222. Plaine, "State Aviation Legislation," p. 336.

223. Plaine, "State Aviation Legislation," p. 336.

224. Federal Airport Act of 1946, Pub. L. No. 377.

225. Federal Airport Act Extension, Pub. L. No. 81-846, 64 Stat. 1071.

226. Airport and Airway Development Act of 1970, Pub. L. No. 91-258, 49 U.S.C. 1730.

227. Federal Aviation Act of 1958, Pub. L. No. 85-726; 72 Stat. 737.

228. *Glatt v. Page,* District Court, 3d Jud. Dist. of Neb., Docket 93-115 (1928).

229. *Glatt v. Page,* Decree of District Court, par. 5.

230. *United States v. Causby,* 328 U.S. 256 (1946).

231. *United States v. Causby*, 328 U.S. at 260–61.

232. 328 U.S. at 264.

233. 328 U.S. at 265.

234. 328 U.S. at 262.

235. 72 Stat. 739, 49 U.S.C. § 1301(24), 49 U.S.C.A. § 1301(24).

236. *Griggs v. Allegheny*, 369 U.S. 84, 85 (1962).

237. 369 U.S. at 89.

238. FAA Modernization and Reform Act of 2012, § 336(a).

239. FAA Reauthorization Act of 2018, Pub. L. No. 115-254.

240. FAA, Flight Advisory: National Special Security Event, Papal Visit, Philadelphia, PA, September 26–27, 2015.

241. For a comprehensive treatment of state drone law, see Laura K. Donohue, "A Tale of Two Sovereigns: Federal and State Use and Regulation of Unmanned Aircraft Systems," in *Handbook of Unmanned Aerial Vehicles*, 2nd ed., ed. Kimon P. Valavanis and George J. Vachtsevanos (New York: Springer International Publishing AG, 2017); National Conference of State Legislatures, "Current Unmanned Aircraft State Law Landscape." For a few examples of state measures see Cal. Civ. Code § 853 (2017); La. Stat. Ann. § 14:63 (2016); Nev. Rev. Stat. Ann. § 493.103(1)

(2015); N.C. Gen. Stat. Ann. § 15A-300.1(b) (2016); Or. Rev. Stat. Ann. § 837.380(1) (2016); Tenn. Code Ann. §§ 39-13-903(a)(1), (3) (2016).

242. See, for example, Cal. Civ. Code § 1708.8(b) (2015); La. Stat. Ann. § 14:63 (2016).

243. Tenn. Code Ann. § 39-14-405 (2016).

244. Nev. Rev. Stat. Ann. § 493.109 (2015).

245. *Bibb v. Navajo Freight Lines Inc.*, 359 U.S. 520, 523 (1959).

246. *South Carolina State Highway Dept. v. Barnwell Bros.*, 303 U.S. 177, 180 (1938).

247. 303 U.S. at 182.

248. 303 U.S. at 184–85.

249. 303 U.S. at 185.

250. 303 U.S. at 187.

251. 303 U.S. at 187.

252. 303 U.S. at 196.

253. 303 U.S. at 196.

254. *Maurer v. Hamilton*, 309 U.S. 589 (1940). Under the 1935 Motor Carrier Act, the Interstate Commerce Commission promulgated regulations to govern the "safety of operation and equipment" of cars in interstate commerce. The regulations failed to address trucks carrying cars over the cab.

255. 309 U.S. at 604–5.

256. 309 U.S. at 605.

257. 309 U.S. at 605.

258. 309 U.S. at 598.

259. *Sproles v. Binford*, 286 U.S. 374, 389–90 (1932), quoting *Morris v. Duby*, 274 U.S. 135, 143 (1927).

260. *Southern Pacific Co. v. Arizona ex rel. Sullivan*, 325 U.S. 761 (1945).

261. *Bibb v. Navajo Freight Lines Inc.*, 359 U.S. 520 (1959).

262. *Skysign International Inc. v. Honolulu*, 276 F.3d 1109, 1113 (2002).

263. 276 F.3d at 1116 (internal quotations and citations omitted).

CHAPTER SIX

1. Jennifer Lynch, "Drone Loans: Customs and Border Protection Records 500 Predator Flights for Other Agencies," *Deeplinks* (blog), Electronic Frontier Foundation, September 27, 2013.

2. See Jake Laperruque and David Janovsky, "These Police Drones Are Watching You," Project on Government Oversight, September 25, 2018.

3. See "Current Unmanned Aircraft State Law Landscape," Transportation, National Conference of State Legislatures, April 1, 2020; see also Amanda Essex,

Taking Off: State Unmanned Aircraft Systems Policies (Washington: National Conference of State Legislatures, 2016).

4. See "Current Unmanned Aircraft State Law Landscape"; see also Essex, *Taking Off: State Unmanned Aircraft Systems Policies.*

5. There are several notable exceptions, such as National Security Letters, 18 U.S.C. §§ 2709, 3511, and FISA Section 702, 50 U.S.C. § 1881a, but these authorities are highly controversial.

6. *United States v. Jones*, 565 U.S. 400, 416 (2011) (Sotomayor, J., concurring) (citation omitted).

7. The key difference between surveillance via electronic means and through traditional human activity is the level of resources required, and, as a result, the scale of surveillance that can occur. Numerous scholars have highlighted that, absent strong legal limits, electronic location tracking exponentially increases the degree to which government can monitor its citizens, and gives it unprecedented power. See Orin S. Kerr, "An Equilibrium-Adjustment Theory of the Fourth Amendment," *Harvard Law Review* 125, no. 2 (2011); see also Kevin S. Bankston and Ashkan Soltani, "Tiny Constables and the Cost of Surveillance: Making Cents Out of *United States v. Jones*," *Yale Law Journal* 123 (2014).

8. *Carpenter v. United States,* 138 S. Ct. 2206, 2218 (2018).

9. Nathan Freed Wessler and Naomi Dwork, "FBI Releases Secret Spy Plane Footage from Freddie Gray Protests," *Speak Freely* (blog), American Civil Liberties Union, August 4, 2016.

10. Shahid Buttar, "Illinois Declines to Adopt Proposed Arbitrary Drone Surveillance of Protests," *Deeplinks* (blog), Electronic Frontier Foundation, June 22, 2018.

11. Harmon Leon, "Top Secret Military-Grade Surveillance Drones Might Be Coming to Your Neighborhood," *Observer,* June 28, 2019; Kyle Mizokami, "Tiny Drone-Based Surveillance System Can Watch over an Entire Small Town," *Popular Mechanics*, March 2, 2017.

12. Laperruque and Janovsky, "These Police Drones Are Watching You."

13. See Jay Stanley, *The Dawn of Robot Surveillance: AI, Video Analytics, and Privacy* (New York: American Civil Liberties Union, 2019).

14. 18 U.S.C. § 2518(3)(c).

As with many surveillance rules and techniques, you can find a great primer on how exhaustion works in *The Wire.* The plot of season 1, episode 4, "Old Cases," largely focuses on detectives engaging in what they perceive as a pointless effort to track targets using traditional tails in order to meet exhaustion requirements for a wiretap order they are seeking.

15. For more information on the electronic exhaustion concept, see Jake Laperruque, "Congress Should Place More Limits on Cellphone Location Tracking after *Carpenter,*" Just Security, March 23, 2018; and Jake Laperruque, "Pri-

vacy after *Carpenter*: We Need Warrants for Real-Time Tracking and 'Electronic Exhaustion,'" Project on Government Oversight, July 2, 2018.

16. See "Federal Bureau of Investigation (FBI)," Martin Luther King, Jr. Encyclopedia, Martin Luther King, Jr. Research and Education Institute, Stanford University; see also Jeffrey O. G. Ogbar, "The FBI's War on Civil Rights Leaders," *Daily Beast*, (originally published January 16, 2017, updated June 3, 2020).

17. See Alvaro M. Bedoya, "The Color of Surveillance," *Slate*, January 18, 2016.

18. Select Committee to Study Governmental Operations with Respect to Intelligence Activities, Final Report, S. Rep. No. 94-755, Book II, U.S. Senate (1976), pp. 5–15 (aka "Church Committee Report").

19. This may seem arduous, but it is unlikely to be if drone use is limited to situations when it is truly necessary, rather than as a generalized surveillance tool. Additionally, new innovations are being made in other sectors of video surveillance to facilitate redactions without overwhelming human effort. See Bill Schrier, "Inside the Seattle Police Hackathon: A Substantial First Step," *GeekWire*, December 20, 2014.

20. 18 U.S.C. § 2518(4).

21. See Kevin Rector, "As Police Weigh Surveillance Program, Private Company at Helm Looks to Court Private Clients," *Baltimore Sun*, October 7, 2016; see also Monte Reel, "Secret Cameras Record Baltimore's Every Move from Above," *Bloomberg Businessweek*, August 23, 2016.

22. Jerry Iannelli, "Miami Beach Police Use Surveillance Blimp in Possible Violation of Florida's Police-Drone Ban," *Miami New Times*, January 9, 2019.

23. Iannelli, "Miami Beach Police Use Surveillance Blimp."

24. See Reel, "Secret Cameras Record Baltimore's Every Move."

25. Notably, sunsets have helped move Congress toward reform of controversial provisions of the USA PATRIOT Act, especially in 2015 when three provisions expired for a short period before Congress responded by passing a major surveillance reform bill, the USA FREEDOM Act.

26. See Laperruque and Janovsky, "These Police Drones Are Watching You"; see also "Eyes in the Sky: Police Use of Drone Technology," panel discussion, Of Rockets and Robotics: The Regulation of Emerging Aerial Technology, a Cato Conference, September 25, 2018, Cato Institute, Washington, MP4 copy, https://cdn.cato.org/archive-2018/cc-09-25-18-04.mp4.

CHAPTER SEVEN

1. On the terminology used for drones, see Jay Stanley, "'Drones' vs. 'UAVs'—What's Behind a Name?," *Free Future* (blog), American Civil Liberties Union (ACLU), May 20, 2013.

2. See, for example, "The End of Privacy: The Surveillance Society," *The Economist*, May 1, 1999, https://web.ics.purdue.edu/~felluga/privacy2.html.

3. Brian Bennett, "Police Employ Predator Drone Spy Planes on Home Front," *Los Angeles Times*, December 10, 2011.

4. Catherine Crump and Jay Stanley, "Why Americans Are Saying No to Domestic Drones," *Slate*, February 11, 2013; Allie Bohm, "Status of 2014 Domestic Drone Legislation in the States," *Free Future* (blog), ACLU, April 22, 2014. Most bills imposed warrant requirements or other restrictions on law enforcement use of the technology.

5. "Operation of Small Unmanned Aircraft Systems over People," Notice of Proposed Rulemaking, FAA-2018-1087, 84 Fed. Reg. 3856 (February 13, 2019); "Operation of Small Unmanned Aircraft Systems over People," Final Rule, FAA-2018-1087, 86 Fed. Reg. 4314 (January 15, 2021).

6. Philip Ross, "FAA Will Let Some Drones Fly Beyond Line of Sight," *IEEE Spectrum*, May 6, 2015.

7. "ARC Recommendations Final Report," UAS Identification and Tracking Aviation Rulemaking Committee, September 30, 2017, p. 6.

8. "Remote Identification of Unmanned Aircraft Systems," Notice of Proposed Rulemaking, FAA-2019-1100, 84 Fed. Reg. 72438, December 31, 2019.

9. Federal Aviation Administration (FAA), *Integration of Civil Unmanned Aircraft Systems in the National Airspace System Roadmap*, 3rd ed. (Washington: FAA, 2020).

10. FAA, *Roadmap*, 2nd ed., p. 4.

11. "AP: FBI Using Low-Flying Spy Planes over U.S.," CBS News, June 2, 2015; Devlin Barrett, "Americans' Cellphones Targeted in Secret U.S. Spy Program," *Wall Street Journal*, November 13, 2014; Andrew Crocker, "New FOIA Documents Confirm FBI Used Dirtboxes on Planes without Any Policies of Legal Guidance," Electronic Frontier Foundation, March 9, 2016.

12. Monte Reel, "Secret Cameras Record Baltimore's Every Move from Above," *Bloomberg Businessweek*, August 23, 2016.

13. Justin Fenton and Talia Richman, "Baltimore Police Back Pilot Program for Surveillance Planes, Reviving Controversial Program," *Baltimore Sun*, December 20, 2019.

14. Arthur Holland Michel, *Eyes in the Sky: The Secret Rise of Gorgon Stare and How It Will Watch Us All* (Boston: Houghton Mifflin Harcourt, 2019), p. 53; Loren Thompson, "Air Force's Secret 'Gorgon Stare' Program Leaves Terrorists Nowhere to Hide," *Forbes*, April 10, 2015.

15. See, for example, L3Harris, "Wide-Area Motion Imagery Intelligence"; Logos Technologies, "Products and Services"; MAG Aerospace, "WAMI (Wide Area Motion Imagery)"; CRI.US, "Products"; Sierra Nevada Corporation, "Sensor and Mission Systems."

16. Ross McNutt, "Wide Area Motion Imagery Technical Overview," Persistent Surveillance Systems, https://1bf873cf-c48d-4ea0-a7e8-b76b71762d31.filesusr.com/ugd/9845f4_93752cf91e1d41cb8388458f0e69f56e.pdf.

17. Fintan Corrigan, "12 Top Lidar Sensors for UAVs, Best Lidar Drones and Great Uses," DroneZon, March 15, 2020; Jay Stanley, "Rapid Improvements in Lidar Technology Could Have Surveillance Implications," *Free Future* (blog), ACLU, February 26, 2014.

18. IPVM, "2021 Camera Book," https://ipvm.com/reports/ip-camera-training-book.

19. "Consumer Video Surveillance Market to Top $1 Billion in 2018, HIS Markit Says," Security Today, November 8, 2018; Sven Skafisk, "This Is How Smartphone Cameras Have Improved over Time," PetaPixel, June 16, 2017.

20. Indigo Vision, "Products, Specialized Cameras," http://web.archive.org/web/20190415230615/https://csmerchants.com.au/pages/indigovision-specialized-cameras.

21. Stephen Shankland, "Canon's 250-Megapixel Sensor Powers Eagle-Eyed Camera," CNET, September 7, 2015.

22. "Rise of the Drones," *NOVA*, PBS. See also Jay Stanley, "Report Details Government's Ability to Analyze Massive Aerial Surveillance Video Streams," *Free Future* (blog), ACLU, April 5, 2013.

23. "Bentley's New 57.7 Gigapixel Ad Is the World's Highest Res Landscape Photo," PetaPixel, March 30, 2017.

24. Quinn Staley, "Surveillance System Helps Officials Identify Security Threats," MIT Lincoln Laboratory, May 6, 2019. See also Department of Homeland Security, "Snapshot: S&T's Immersive Imaging System's High-Resolution Images and 360-Degree Coverage Provides Full Scene Situational Awareness," Science and Technology, April 30, 2019.

25. Tyler Rogoway and Joseph Trevithick, "This Mysterious Military Spy Plane Has Been Flying Circles over Seattle for Days," *The War Zone* (blog), The Drive, August 3, 2017.

26. Amanda Berg, Jörgen Ahlberg, and Michael Felsberg, "Channel Coded Distribution Field Tracking for Thermal Infrared Imagery," Computer Vision Foundation, 2016.

27. Daniel König et al., "Fully Convolutional Region Proposal Networks for Multispectral Person Detection," Computer Vision Foundation, 2017. Multispectral drone cameras are already broadly available and are especially popular for agricultural use. See Fintan Corrigan, "Multispectral Imaging Camera Drones in Farming Yield Big Benefits," DroneZon, February 10, 2020.

28. Chen Chen et al., "Learning to See in the Dark," Computer Vision Foundation, 2018, http://openaccess.thecvf.com/content_cvpr_2018/papers/Chen_Learning_to_See_CVPR_2018_paper.pdf.

29. "What Is Super-Resolution?," Unique Solutions, Infognition.

30. Michael Hicks, "The Secret Behind 8K Upscaling," TV Tech, October 8, 2019.

31. See Jay Stanley, *The Dawn of Robot Surveillance: AI, Video Analytics, and Privacy* (Washington: ACLU, June 2019). I have drawn from this report here.

32. Mahdieh Poostchi et al., "Semantic Depth Map Fusion for Moving Vehicle Detection in Aerial Video," Computer Vision Foundation, 2016 (internal citations omitted). See also Riad Hammoud et al., "Automatic Association of Chats and Video Tracks for Activity Learning and Recognition in Aerial Video Surveillance," Research Gate, October 2014.

33. Mohammadamin Barekatain et al., "Okutama-Action: An Aerial View Video Dataset for Concurrent Human Action Detection," Computer Vision Foundation, 2017; Gui-Song Xia et al., "DOTA: A Large-scale Dataset for Object Detection in Aerial Images," Computer Vision Foundation, 2018; Alexandre Alahi, Vignesh Ramanathan, and Li Fei-Fei, "Socially-Aware Large-Scale Crowd Forecasting," Stanford University, 2014.

34. "DIUx xVIEW 2018 Detection Challenge," Defense Innovation Unit Experimental, 2018, http://xviewdataset.org/.

35. Rui Li et al., "Monocular Long-term Target Following on UAVs," Proceedings of the IEE Conference on Computer Vision and Pattern Recognition Workshops (2016): 29-37, https://openaccess.thecvf.com/content_cvpr_2016_workshops/w3/papers/Li_Monocular_Long-Term_Target_CVPR_2016_paper.pdf; IPSOTEK, "Investigation and Forensics"; Barekatain et al., "Okutama-Action: An Aerial View Video Dataset"; Burak Uzkent, Matthew J. Hoffman, and Anthony Vodacek, "Real-Time Vehicle Tracking in Aerial Video Using Hyperspectral Features," Computer Vision Foundation, 2016, http://openaccess.thecvf.com/content_cvpr_2016_workshops/w29/papers/Uzkent_Real-Time_Vehicle_Tracking_CVPR_2016_paper.pdf ("Aerial vehicle detection and tracking has attracted considerable interest in the computer vision community"); Poostchi et al., "Semantic Depth Map Fusion" ("The ultimate goal of our system is to achieve highly reliable motion detection to perform persistent tracking of moving vehicles over long time frames in large scale urban imagery.").

36. Michel, *Eyes in the Sky*, pp. 124–34; Arthur Holland Michel, "How Big Tech Is Helping Build the Pentagon's All-Seeing Eye-in-the-Sky," Fast Company, June 18, 2019.

37. Dell Cameron and Kate Conger, "Google Is Helping the Pentagon Build AI for Drones," *Gizmodo*, March 6, 2018; Kate Conger, "Google Plans Not to Renew Its Contract for Project Maven, a Controversial Pentagon Drone AI Imaging Program," *Gizmodo*, June 1, 2018.

38. Becky Peterson, "Palantir Took Over from Google on Project Maven," *Business Insider*, December 10, 2019.

39. *United States v. Maynard,* 615 F.3d 544, 562 (D.C. Cir. 2010).

40. Philippe Golle and Kurt Partridge, "On the Anonymity of Home/Work Location Pairs," in *Pervasive '09: Proceedings of the 7th International Conference on Pervasive Computing*, ed. Hideyuki Tokuda and Michael Beigl (Berlin: Springer-Verlag, 2009), http://crypto.stanford.edu/~pgolle/papers/commute.pdf.

41. See, for example, John Krumm, "Inference Attacks on Location Tracks," in *Pervasive '07: Proceedings of the 5th International Conference on Pervasive Computing*, Toronto, May 2007, www.research.microsoft.com/en-us/um/people/jckrumm/Publications%202007/inference%20attack%20refined02%20distribute.pdf.

42. Jianjun Gao et al., "Pattern of Life from WAMI Objects Tracking Based on Visual Context-Aware Tracking and Infusion Network Models," *Proceedings of SPIE* 8745 (May 2013); Maxar Technologies, "Discovering Pattern of Life Activity Using Machine Learning," *Earth Intelligence* (blog), Maxar, March 14, 2018; Michel, *Eyes in the Sky*, pp. 140, 162.

43. Michel, *Eyes in the Sky*, p. 133; Steven Thomas Smith, "Network Discovery Using Wide-Area Surveillance Data," Research Gate, August 2011.

44. Alahi et al., "Socially-Aware Large-Scale Crowd Forecasting."

45. Chaoming Song et al., "Limits of Predictability in Human Mobility," *Science* 327 (February 2010): 1018, http://zehui.yolasite.com/resources/Limits%20of%20Predictability%20in%20Human%20Mobility.pdf.

46. Alahi et al., "Socially-Aware Large-Scale Crowd Forecasting."

47. Joseph Flynt, "Autonomous Drones: 7 Best Autonomous Self-Flying Drones," 3D Insider, December 24, 2018.

48. Colin Snow, "Sense and Avoid for Drones Is No Easy Feat," Drone Analyst, September 22, 2016; JWalters, "Sense and Avoid for Small Unmanned Aerial Systems," *An Unmanned Future* (blog), January 10, 2018.

49. Brian Garrett-Glaser, "Iris to Test Commercial Drone Sense and Avoid System in BVLOS First," Aviation Today, August 15, 2019; BIS Research, "A Major Growth Area in the Drone Market: Sense and Avoid Technology," Market Research, May 9, 2018.

50. David Martin, "New Generation of Drones Set to Revolutionize Warfare," *60 Minutes*, CBS News, January 8, 2017; and Amy McCullough, "The Looming Swarm," *Air Force Magazine*, March 22, 2019.

51. Aaron Boyd, "The Pentagon Wants AI-Driven Drone Swarms for Search and Rescue Ops," Nextgov, December 26, 2019.

52. Michel, *Eyes in the Sky*, p. 306.

53. 88th Air Base Wing Public Affairs, "AFRL Successfully Completes Two and a Half–Day Flight of Ultra Long Endurance Unmanned Air Platform (LEAP)," Air Force Materiel Command, December 12, 2019.

54. Alexander M. G. Walan, "Adaptable Lighter Than Air (ALTA)," Defense Advanced Research Projects Agency.

55. Spencer Ackerman, "Homeland Security Wants to Spy on 4 Square Miles at Once," *Wired*, January 23, 2012; Spencer Ackerman, "DHS Uses Wartime Mega-Camera to Watch Border," *Wired*, April 2, 2012; Michel, *Eyes in the Sky*, p. 57.

56. Jamie Schwandt, "It's Time to Get Yourself a Tethered Drone," *The Long March* (blog), Task and Purpose, October 8, 2018. See also, for example, Power-Line, "Tethered Drones/Unlimited Power"; ElistAir, https://elistair.com/; Hover-fly, "Technology Overview Video."

57. "An Airbus Drone Has Set the Record for the Longest Continuous Flight within Earth's Atmosphere," MIT Technology Review, August 8, 2018.

58. Dricus De Rooij, "Top 8 Solar Powered Drone (UAV) Developing Companies," Sino Voltaics, July 9, 2015.

59. Mike Ball, "UAVOS Stratospheric UAV Reaches 30km Altitude," Unmanned Systems News, Unmanned Systems Technology, August 30, 2019.

60. Wikipedia, "Lockheed-Martin High-Altitude Airship," Atmospheric Satellite.

61. See, for example, Jay Stanley, "The Use of Killer Robots by Police," *Free Future* (blog), ACLU, December 20, 2016.

62. *Tennessee v. Garner,* 471 U.S. 1 (1985).

63. Jay Stanley, "Five Reasons Armed Domestic Drones Are a Terrible Idea," *Free Future* (blog), ACLU, August 27, 2015.

64. Aviation Committee, "Recommended Guidelines for the Use of Unmanned Aircraft," International Association of Chiefs of Police, August 2012.

65. "Groups Concerned over Arming of Domestic Drones," CBS DC, May 23, 2012.

66. Philip Bump, "The Border Patrol Wants to Arm Drones," *The Atlantic*, July 2, 2013.

67. Stanley, "Five Reasons Armed Domestic Drones Are a Terrible Idea."

68. Kelsey D. Atherton, "The Pentagon Is Flying More Drone Missions along America's Border," *Defense News*, February 10, 2019.

69. Office of Inspector General, *CBP Has Not Ensured Safeguards for Data Collected Using Unmanned Aircraft Systems*, OIG-18-79 (Washington: Department of Homeland Security, September 21, 2018), pp. 4, 7. The Texas center is the base for some manned flights and those may be included in the 635 number.

70. Military Factory, "General Atomics MQ-9 Reaper (Predator B): All-Weather Unmanned Combat Aerial Vehicle."

71. Department of Homeland Security (DHS), *Privacy Impact Assessment for the Aircraft Systems*, DHS/CBP/PIA-018 (Washington: DHS, September 9, 2013), p. 7.

72. Office of Inspector General, *CBP Has Not Ensured Safeguards*, p. 6.

73. Office of Inspector General, *CBP Has Not Ensured Safeguards*, p. 6.

74. U.S. Government Accountability Office, "Unmanned Aerial Systems: Department of Homeland Security's Review of U.S. Customs and Border Protection's Use and Compliance with Privacy and Civil Liberty Laws and Standards," Briefing for Staff of the Subcommittees on Homeland Security, U.S. Senate and House Committees on Appropriations, September 30, 2014, https://www.gao.gov/assets/670/666282.pdf.

75. Sandia National Laboratories, "What Is Synthetic Aperture Radar (SAR)?"

76. William Welsh, "VADER Radar to Thwart Roadside Bombers," Defense Systems, April 27, 2010.

77. Patrick C. Miller, "UAV Sensor Sensibility," *UAS Magazine*, April 18, 2016.

78. John Keller, "Northrup to Operate Man-Hunting Airborne Radar Systems for Operations in Afghanistan," Military and Aerospace Electronics, December 16, 2013.

79. Miller, "UAV Sensor Sensibility."

80. Department of Homeland Security, *Privacy Impact Assessment for the Aircraft Systems*, p. 8.

81. David Bier and Matthew Feeney, "Drones on the Border: Efficacy and Privacy Implications," Immigration Research and Policy Brief no. 5, Cato Institute, May 1, 2018.

82. U.S. Customs and Border Protection, "CBP to Test the Operational Use of Small Unmanned Aircraft Systems in 3 U.S. Border Patrol Sectors," media release, September 14, 2017; Nasdaq "AeroVironment Receives $5.25 Million Puma 3 AE Contract for U.S. Border Control," press release, October 22, 2019, https://www.nasdaq.com/press-release/aerovironment-receives-%245.25-million-puma-3-ae-contract-for-u.s.-border-patrol-2019.

83. Department of Homeland Security, *Privacy Impact Assessment for the Aircraft Systems*, p. 3.

84. Shirin Ghaffary, "The 'Smarter' Wall: How Drones, Sensors, and AI Are Patrolling the Border," Recode, *Vox*, February 7, 2020.

85. Department of Homeland Security, *Privacy Impact Assessment for the Aircraft Systems*, pp. 2, 4.

86. Ghaffary, "The 'Smarter' Wall." See also "AeroVironment Receives $5.25 Million Puma 3 AE Contract for U.S. Border Patrol."

87. Department of Homeland Security, "DHS S&T Awards $200K to San Diego's Planck Aerosystems Inc. for Final Testing of Small Unmanned Aircraft System," news release, Science and Technology, August 29, 2019.

88. Ghaffary, "The 'Smarter' Wall."

89. Michel, *Eyes in the Sky*, p. 82.

90. General Atomics Aeronautical Systems Inc., https://www.ga-asi.com/images/products/aircraft_systems/pdf/MQ-9B-Capability-Profile-II.pdf.

91. John Keller, "Northrop to Operate Man-Hunting Airborne Radar Operations in Afghanistan," Military and Aerospace Electronics, December 16, 2013.

92. Ackerman, "Homeland Security Wants to Spy on 4 Square Miles"; Ackerman, "DHS Uses Wartime Mega-Camera"; Michel, *Eyes in the Sky*, p. 57.

93. Dave Long, "CBP's Eyes in the Sky: CBP's Tethered Aerostats Keep Watch for Trouble from 10,000 Feet," *Frontline*, December 2015, CBP (emphasis added).

94. Army Research Laboratory, *History of the U.S. Army Research Laboratory* (Washington: Department of the Army, September 2017), p. 73. See also Long, "CBP's Eyes in the Sky"; "Sentinels of the Sky: The Persistent Threat Detection System," Lockheed Martin, 2018.

95. Michel, *Eyes in the Sky*, p. 83.

96. Department of Homeland Security, *Privacy Impact Assessment for the Aircraft Systems*, p. 8.

97. Randolph Alles, Assistant Commissioner, CBP Office of Air and Marine; Mark Borkowski, Assistant Commissioner, CBP Office of Technology Innovation and Acquisition; and Ron Vitiello, Deputy Chief, CBP Office of Border Patrol, Testimony before the Senate Committee on Homeland Security and Governmental Affairs, hearing on Securing the Border: Fencing, Infrastructure, and Technology Force Multipliers, 114th Cong., 1st sess., May 13, 2015.

98. See multiple articles at ACLU, "Privacy at Borders and Checkpoints," News and Commentary.

99. Office of Inspector General, *CBP Has Not Ensured Safeguards*, p. 10.

100. Russell Brandom, "The US Border Patrol Is Trying to Build Face-Reading Drones," *The Verge*, April 6, 2017.

101. Jennifer Lynch, "Customs and Border Protection Logged Eight-Fold Increase in Drone Surveillance for Other Agencies," Electronic Frontier Foundation, July 3, 2013; Caroline Cournoyer, "Police Agencies Using Border Patrol's Drones More Often Than Thought," Governing, January 15, 2014; Brian Bennett, "Police Employ Predator Drone Spy Planes on Home Front," *Los Angeles Times*, December 10, 2011.

102. Rogoway and Trevithick, "Mysterious Military Spy Plane Has Been Flying Circles over Seattle."

103. Arthur Holland Michel, "The Military-Style Surveillance Technology Being Tested in American Cities," *The Atlantic*, August 3, 2019; Sensor Data Management System, "Columbus Large Image Format (CLIF) 2007 Dataset Overview," https://www.sdms.afrl.af.mil/index.php?collection=clif2007.

104. Mark Harris, "Pentagon Testing Mass Surveillance Balloons across the US," *The Guardian*, August 2, 2019.

105. Wikipedia, "Posse Comitatus Act."

106. Gregg Zoroya and Alan Gomez, "Pentagon Has a 'Unique' Policy for Legal Use of Drones in U.S.," *USA Today*, March 9, 2016.

107. See Gregg Zoroya, "Pentagon Report Justifies Deployment of Military Spy Drones over the U.S.," Opinion, *USA Today*, March 9, 2016.

108. Atherton, "Pentagon Is Flying More Drone Missions."

109. DoD UAS Operations/Exercises 2017–2018, https://dod.defense.gov/Portals/1/Documents/PDFs/FY18 DoD UAS Domestic Use.pdf. See also DoD UAS Operations/Exercises 2011–2017, https://dod.defense.gov/Portals/1/Documents/Web%20site%20UAS%20Tracker.pdf; Dan Gettinger, "Documents: DoD's Domestic Drone Missions," Center for Study of the Drone at Bard College, January 21, 2019.

110. James Laporta, "Military Drones and Surveillance Planes Remain an Option for U.S.-Mexico Border, Documents Show," *Newsweek*, August 23, 2019. See also Michael Shear and Thomas Gibbons-Neff, "Trump Sending 5,200 Troops to the Border in an Election-Season Response to Migrants," *New York Times*, October 29, 2018.

111. Valerie Insinna, "Unmanned Aircraft Could Provide Low-Cost Boost for Air Force's Future Aircraft Inventory, New Study Says," *Defense News*, October 29, 2019.

112. Craig Whitlock, "Crashes Mount as Military Flies More Drones in U.S.," *Washington Post*, June 22, 2014.

113. Dan Gettinger, "Summary of Drone Spending in the FY 2019 Defense Budget Request," Center for the Study of the Drone at Bard College, April 2018.

114. FAA Modernization and Reform Act of 2012, Pub. L. No. 112-95 (February 14, 2012).

115. U.S. Government Accountability Office (GAO), *Unmanned Aircraft Systems: FAA Could Better Leverage Test Site Program to Advance Drone Integration*, GAO-20-97 (Washington: GAO, January 2020), p. 5.

116. Chris Cole, *Accidents Will Happen*, (Oxford, UK: Drone Wars UK, June 27, 2019).

117. Whitlock, "Crashes Mount as Military Flies More Drones." See also Jay Stanley, "Drones, Accidents, and Secrecy," *Free Future* (blog), ACLU, July 2, 2014.

118. Jay Stanley, "The President Reads His Daily Brief on an iPad (and Other Lessons from the NSA)," *Free Future* (blog), ACLU, September 14, 2012.

119. FAA, *Roadmap*, 2nd ed.

120. Barry Summers, "Three Air Force Colonels and the Opening of U.S. Airspace to Military Drones," Know Drones, October 17, 2019.

121. Duncan Hunter National Defense Authorization Act for Fiscal Year 2009, Pub. L. No. 110-417 (October 14, 2008).

122. National Defense Authorization Act for Fiscal Year 2010, Pub. L. No. 111-84, Sec. 935(a).

123. Department of Defense, *Final Report to Congress on Access to National Airspace for Unmanned Aircraft Systems* (Washington: DOD, October 2010).

124. FAA, *Roadmap*, 2nd ed.

125. Department of Defense, *Report to Congress on Future Unmanned Aircraft Systems: Training, Operations, and Sustainability* (Washington: DOD, April 2012).

126. GAO, *FAA Could Better Leverage Test Site Program*, p. 11.

127. Secretary of Defense, Guidance for the Domestic Use of Unmanned Aircraft Systems in U.S. National Airspace, memorandum, August 18, 2018, cited in Gettinger, "Documents: DOD's Domestic Drone Missions."

128. Robert Mueller, FBI Director, Testimony before the Senate Judiciary Committee, 113th Cong., 1st sess., June 19, 2013, https://www.judiciary.senate.gov/meetings/oversight-of-the-federal-bureau-of-investigation-2013-06-19; Dan Roberts, "FBI Admits to Using Surveillance Drones over US Soil," *The Guardian*, June 19, 2013.

129. FAA, *Roadmap*, 2nd ed., pp. 21, 29.

130. Vicki Speed, "FAA Requests for Emergency UAS Airspace Authorizations on the Rise," Inside Unmanned Systems, August 27, 2019.

131. FAA, "Emergency Situations," Unmanned Aircraft Systems Advanced Operations, August 27, 2020; Doug Bonderud, "Agencies Fly Drones to Survey Post-Disaster Damage," Fed Tech Hardware, November 27, 2019.

132. Speed, "FAA Requests for Emergency UAS Airspace Authorizations."

133. Kate Baggaley, "Drones Are Fighting Wildfires in Some Very Surprising Ways," NBC News, November 16, 2017; U.S. Forest Service, "Unmanned Aircraft Systems (UAS)," Managing the Land (undated).

134. Jack Gillum, "Ferguson No-Fly Zone Aimed at Media," AP, November 2, 2014; Max Fisher, "If Police in Ferguson Treat Journalists Like This, Imagine How They Treat Residents," *Vox*, August 18, 2014.

135. Anthony Rothert and Lee Rowland, letter to Reggie Govan, Chief Counsel, FAA, ACLU Legal Department, November 4, 2014.

136. Anthony Rothert and Lee Rowland, letter to Reggie Govan, Chief Counsel, FAA, ACLU Legal Department, December 16, 2016, https://www.aclu.org/sites/default/files/field_document/faa_letter_for_standing_rock_12.16.2016.pdf; Ben Norton, "Dakota Pipeline Protesters Say They Were Detained in Dog Kennels; 268 Arrested in Week of Police Crackdown," *Salon*, November 1, 2016.

137. Greg Timberg, "Surveillance Planes Spotted in the Sky for Days after West Baltimore Rioting," *Washington Post*, May 5, 2015.

138. Eric Tucker, "Comey: FBI Used Aerial Surveillance above Ferguson," AP, October 22, 2015.

139. "AP: FBI Using Low-Flying Spy Planes over U.S.," CBS News, June 2, 2015.

140. Nathan Freed Wessler, "FBI Documents Reveal New Information on Baltimore Surveillance Flights," *Free Future* (blog), ACLU, October 30, 2015; FLIR, "Talon," https://flir.netx.net/file/asset/11061/original.

141. Nathan Freed Wessler, "FBI Releases Secret Spy Plane Footage of Freddie Gray Protests," *Speak Freely* (blog), ACLU, August 4, 2016.

142. Peter Aldhous and Charles Seife, "See Maps Showing Where FBI Planes Are Watching from Above," Spies in the Skies, BuzzFeed News, April 6, 2016.

143. Office of the Inspector General, *Audit of the Department of Justice's Use and Support of Unmanned Aircraft Systems* (Washington: Department of Justice, March 2015), pp. ii, 3.

144. "Downed Drones: ATF Spent $600K on 11 Drones That Never Flew, Report Says," NBC News, March 26, 2015.

145. Office of Public Affairs, "Department of Justice Announces Update to Policy on Use of Unmanned Aircraft Systems," Justice News, November 27, 2019.

146. "Department of Justice Policy on the Use of Unmanned Aircraft Systems," Title 9-95.100, Justice Manual, Department of Justice.

INDEX

Note: Page numbers with "n" indicate endnotes; page numbers with "f" indicate figures.

AAM. *See* advanced air mobility (AAM)
Abbreviated Injury Scale, 22
above ground level (AGL), 56, 67–68, 70
Academy of Model Aeronautics, 11–12, 54
access to airspace, 111–15, 120–25
accountability, 107–8
Accursius, Franciscus, 131–32
adaptability, 94, 105
ad coelum, 69, 118, 131–42, 147, 148, 150, 153–56, 159, 160, 163, 236n22
administrative allocation, 115–16
administrative assignment, 120–25, 130
administrative mandates, 93, 103
Administrative Procedure Act, 25, 212–13n32
admiralty clause, 148
advanced air mobility (AAM), 47
Advance Notice of Proposed Rulemaking (ANPRM), 46
Advisory Circular on Model Aircraft Operating Standards, 54
Aeneid (Virgil), 131, 244n3
aerial corridors, 96–99, 112–14, 127, 129–30
aerial imagery, 188–90. *See also* photography
aerial surveillance. *See* surveillance
aerial trespass, 39–40, 68–72
aerostats, 192, 197
AGL. *See* above ground level (AGL)
agriculture, 58, 76–77
AI. *See* artificial intelligence (AI)
Airbus, 114, 192, 266n57
air carriers, 15–16, 36–37, 46, 213n36, 226n24, 243n101
Air Commerce Act of 1926, 119, 150–56, 239n39, 239n42, 254n158, 257n210
aircraft certification, 7, 36, 64–65, 85
aircraft-launched drone swarms, 191
Air Force Special Operations Command, 199
air highways, 96–98, 232n83
Airline Deregulation Act of 1978, 16, 68, 243n101
air lots, 118
Air Mail Act of 1928, 121
Air Mail Act of 1934, 122
airmail routes, 120–22
AirMap, 31–32
air navigation easements, 98–99
Airport and Airway Development Act of 1970, 158
airport perimeter monitoring, 58
airport slots, 123–24
air rights, 98
Air Safety Board, 157

airspace as property, 72, 117–20
airspace authorizations, 30–32
air taxi operators, 371
air traffic control (ATC), 31, 86, 125–26
air traffic facilities, 31–32
air traffic management (ATM), 183
air transportation, 15–16
airworthiness certification, 10–11, 36–37, 54–55, 61–62, 64
Alabama, 145
ALI. *See* American Law Institute (ALI)
Alito, Samuel, 102
allocation, 115–16
Alphabet, 36, 79
Amazon, 79
Amazon Prime Air, 62
American Airways, 122
American Bar Association, 147–48, 253n133
American Civil Liberties Union, 205
American colonies, 138–39
American Law Institute (ALI), 40, 69–70, 72
American Society for Testing and Materials (ASTM), 63–64
Anglo-American law, 117–18, 131–32
ANPRM. *See* Advance Notice of Proposed Rulemaking (ANPRM)
anticipatory regulation, 94
ARC. *See* Aviation Rulemaking Committee (ARC)
architecture, 77–78
Arizona, 104, 106
armed drones, 192–93. *See also* weapons
Army Air Corps, 122
Articles of Confederation, 143
artificial intelligence (AI), 182, 187–89
assignment of airspace, 120–25, 130
Associated Press, 205
ASTM. *See* American Society for Testing and Materials (ASTM)
ATC. *See* air traffic control (ATC)
ATF. *See* Bureau of Alcohol, Tobacco, Firearms and Explosives (ATF)
ATM. *See* air traffic management (ATM)
atmospheric satellites, 192
AT&T, 83, 231n51
auctioning slots, 124
automation, 114–15, 125, 178, 187–90
autonomous flight, 190–91
aviation law, 147–48
Aviation Rulemaking Committee (ARC)
  Final Report, 215n66
  Micro UAS ARC, 21–22, 44
  Remote ID and Tracking ARC, 23–27, 215n73
  Small UAS ARC, 21–22
aviation safety, 5, 7, 38–39, 156–58, 216n83
avigation easements, 98–99, 109, 243n100

Baecher, Gregory, 95–96
balloons, 191–92, 200
Baltimore, Maryland, 205
Baltimore Police Department, 184, 187
Banjo, 232–33n93
Baten's Case, 133, 135, 245n17
battery life, 169
behavior of drones, 190–92
BEYOND program, 29–30, 40, 103, 217n92
beyond the visual line of sight (BVLOS)
  detect-and-avoid (DAA), 64, 182, 191
  domestic military flights, 201, 203
  domestic use of drones, 182
  expanded operations, 46
  Part 107, 18, 60–61, 65, 67
  remote ID, 41, 43–44
  Section 91, 137, 225–26n21
  Section 44807, 62, 65
  state and local governments, 67
Bezos, Jeff, 79
Bingham, Hiram, 239n41
Blackburn, Sir Colin, 135
Blackstone, Sir William, 133–34
blimps, 192
Bloomberg Businessweek, 184
Boeing, 192
border drones, 194–99
Braniff Airways v. Neb. State Board of Equalization & Assessment, 163, 239n40, 239n43, 257n211
Brennan, William, 1–2
broadcast remote ID, 41–44, 49, 63
Brown, Walter, 121–22
Bureau of Air Commerce, 157
Bureau of Alcohol, Tobacco, Firearms and Explosives (ATF), 199, 206, 271n144
Bureau of Lighthouses, 254n161
Bureau of Safety Regulations, 158
Bush, George W., 81, 124
Butler v. Frontier Telephone Company, 141
buyer beware doctrine, 107
BuzzFeed, 205
BVLOS. *See* beyond the visual line of sight (BVLOS)

CAA. *See* Civil Aeronautics Authority (CAA)
California, 138–39
Callaway, Llewellyn, 140
camera technology, 55–56, 183–87
cargo shipping, 80
Carpenter v. United States, 168
Carus, Titus Lucretius, 131, 244n2
case-by-case authorization, 13–14
Category 1 UAS, 22, 44–46
Category 2 UAS, 215n66
Category 3 UAS, 215n66
Category 4 UAS, 22, 45–46
C-band spectrum, 50
CBP. *See* U.S. Customs and Border Protection (CBP)
cellphone tracking, 168
cell-site simulators, 183, 205
cells on wheels (COWs), 83
census blocks, 189
Certificates of Waiver or Authorization (COAs), 11, 13, 14, 54–55, 58
Chicago, 118
Church, Frank, 173
Church Committee, 172–73
Civil Aeronautics Act of 1938, 122, 156–58
Civil Aeronautics Authority (CAA), 157–58
Civil Aeronautics Board, 122
civil liability, 99
civil operation, 10–11
clearance procedures, 31
Clear and Present Safety (Zenko), 81
Clifton v. Bury, 136
coal mining, 141
COAs. *See* Certificates of Waiver or Authorization (COAs)
Code of Virginia 1860, 247n55
COINTELPRO. *See* counterintelligence program (COINTELPRO)
Coke, Edward, 133
Coltman, Sir Thomas, 135
Comey, James, 205
Commentaries on American Law (Kent), 138, 246–47n52
Commentaries on the Law of England (Blackstone), 133–34, 236n22, 245n18–20
Commerce Clause, 145–46, 151–52
Commerce Department, 151, 228n57
commercial operations
  air highways, 96–98
  avigation easements, 99
  cooperative federalism, 126
  expanded operations, 19–20, 34
  nighttime operations, 46
  no-fly provision, 68
  Part 107, 59–61, 85
  pilot certifications, 53
  regulatory hurdles, 85
  remote ID, 41–44
  Section 333, 12–13, 56–58
  state and local governments, 102, 161
  technology advancement, 111–12
common carriage, 213n36
Commonwealth of Massachusetts, 138
Commonwealth of Pennsylvania, 141
Commonwealth v. Young, 144–45
communications, 182–83
condominium laws, 118
congestion, 47, 80, 120, 123, 125
Congress
  administrative assignment of airspace, 120–22
  Air Commerce Act of 1926, 150–52
  air traffic control (ATC), 86
  airworthiness certification, 64
  border drones, 199
  Civil Aeronautics Act of 1938, 157
  emergency services, 59
  FAA creation, 85
  FAA Modernization and Reform Act of 2012, 56–57
  FAA Reauthorization Act of 2018, 33, 96, 225n17
  federal preemption, 68, 129
  federal-state divide, 113, 117–20
  navigable airspace, 159–60
  property rights, 143–46
  reform bills, 261n25
  sandboxes, 105
  Section 44807, 65
  surface airspace, 117–20
  UAS integration, 9–10, 56, 112, 201–2
  UAS registration, 27
Congressional Research Service, 113
Constitution, 143
construction industry, 77–78
consumer drones, 190–91
Consumer Electronics Show, 22
consumer protection, 105–8. *See also* Market Empowerment Framework
counterintelligence program (COINTELPRO), 173
cooperative federalism, 126–29
Corbett v. Hill, 135, 246n29

coronavirus pandemic, 60–61, 78–79, 82–83
counterintelligence program (COINTELPRO), 173
counter-UAS authority, 34–35, 226n28
Cowen, Tyler, 111
COWs. *See* cells on wheels (COWs)
Cresswell, Sir Cresswell, 135
critical infrastructure, 47–48
crony capitalism, 106–7
crop monitoring, 76–77
cross-governmental UAS Executive Committee, 202
Curtiss Airports, 153
Curtiss-Wright subsidiary, 153

DAA. *See* detect-and-avoid (DAA)
Da-Jiang Innovations (DJI), 83, 95
Dakota Access Pipeline, 204–5
D'Andrea, Raffaello, 78
de caelo servare and *de caelo tactus,* 131
deep learning, 187
Defense Advanced Research Projects Agency (DARPA), 192
defense and security, 81–82. *See also* U.S. military
delivery, 16, 36–37, 47, 60–61, 78–80, 99, 112, 115
Department of Commerce and Labor, 255n161
Department of Defense, 35
Department of Energy, 35
Department of Homeland Security (DHS), 1, 35, 165, 186, 194
departments of transportation (DOTs), 127. *See also* U.S. Department of Transportation (DOT)
De Rerum Natura (Carus), 131, 244n2
detect and avoid (DAA), 63–64, 182, 191
DHS. *See* Department of Homeland Security (DHS)
dirtboxes, 183–84
DJI. *See* Da-Jiang Innovations (DJI)
DOJ. *See* U.S. Department of Justice (DOJ)
domestic drones, 8–9, 181–82, 193–94, 200–203
domestic military flights, 199–203
DOT. *See* U.S. Department of Transportation (DOT)
Douglas, William O., 69
Drone Advisory Committee (DAC), 224n1
droneports, 128–29
drones. *See also* commercial operations; integration into national airspace; law enforcement agencies; surveillance
- aerial corridors, 96–98, 112–14, 127, 129–30
- agriculture, 58, 76–77
- air carriers and, 15–16, 36–37, 46, 213n36, 226n24, 243n101
- as aircraft, 55–56, 85
- aircraft certification, 7, 36, 64–65, 85
- air rights, 98
- airspace as property, 72, 117–20
- architecture, 77–78
- armed, 192–93
- automation, 114–15, 125, 178, 187–90
- avigation easements, 98–99
- behavior of, 190–92
- border, 194–99
- camera technology, 55–56, 183–87
- cargo shipping, 80
- construction, 77–78
- consumer, 190–91
- defense and security, 54–55, 69, 81–82, 100, 119, 165, 184–91, 199–203
- delivery, 16, 36–37, 47, 60–61, 78–80, 99, 112, 115
- detect-and-avoid (DAA), 63–64, 182, 191
- development of operations, 54–55
- domestic, 8–9, 181–82, 193–94, 200–203
- emergencies and, 59, 82–84, 175–77, 204–5
- enabling legislation, 8–12
- expanded operations, 19–20, 27–30, 44–47, 181–83
- federal-administrative model, 115, 129–30
- government use of, 99–102
- highway use, 127–28
- infrastructure and, 47–48, 80
- lawsuits, 98, 99, 128, 243n100
- line of sight requirements, 94–96
- long-distance operations, 114–15
- Market Empowerment Framework, 88–94, 100
- media and entertainment, 84
- medical supplies, 60–61, 83
- models of airspace regulation, 115–16
- nighttime operations, 18, 44, 46, 182
- nomenclature, 224n1
- nuisance, 72–73, 134–36, 139–40, 142, 150, 153–55, 158–60

number of, 53
passenger, 114, 243–44n104
permissionless innovation, 87–88, 91
policy issues, 85–86
privacy, 73, 82, 101–2, 170–72, 181, 198–99, 206, 228n57
public acceptance of, 66
registration requirements, 25–27, 39, 43, 57, 67, 216n83
regulatory progress, 5–8
sandboxes, 91, 102–8
solar-powered, 192
spectrum and, 49–51
state and local government, 38–41, 65–68
state-market model, 116, 126–29
technology advancement, 111–12, 182–83
trespass and, 39–40, 68–72, 99, 133–37, 139–42, 149–50, 153–56, 158–60
use cases of technology, 76–84
waiver process, 17–19, 59–61
as weapon, 81, 100, 193
Drone Tort Committee, 70–72, 227n52
DroneUp, 112
dual sovereignty, 143, 151

Eastern Air Transport, 122
Electronic Frontier Foundation, 193, 199
electronic signals collection, 183
electronic surveillance, 167–68, 171, 260n7
Ellenborough, Lord, 134–35, 137
emergency services, 59, 82–84, 175–77, 204–5
English common law, 131–39, 146, 148, 149–50
Entick v. Carrington, 136–37
entrepreneurs, 100
exhaustion requirements, 165–66, 170–72, 260n14
expanded operations, 19–20, 27–30, 44–47, 181–83
experimentation, 91
exposure control, 107–8

FAA. *See* Federal Aviation Administration (FAA)
FAA Authorization Act of 2018, 96
FAA Extension, Safety, and Security Act (2016 Extension Act), 23, 47, 48, 63
FAA Modernization and Reform Act (FMRA), 9–12, 24–25, 34, 55–57, 105, 201, 212n13
FAA Office of the Chief Counsel, 38
FAA Reauthorization Act of 2018, 32–35, 33f, 46, 48–50, 61, 65, 218n109, 219n128, 225n17
FAA-recognized identification areas (FRIAs), 42–44
FAA's 2015 Fact Sheet, 67
FAA UAS Symposium, 60, 73–74
face recognition, 169, 199
Fay v. Prentice, 135
FBI, 203–6
FCC. *See* Federal Communications Commission (FCC)
federal-administrative model, 115, 129–30
Federal Airport Act, 158
Federal Air Service, 148
Federal Aviation Act, 15, 67–68, 73, 120
Federal Aviation Administration (FAA). See also Part 107
ad coelum, 132
aerial corridors, 112–13
air carriers, 15–16, 36–37, 46, 226n24, 243n101
air highways, 96–98
airport slots, 123–24
airworthiness certification, 10–11, 36–37, 54–55, 61–62, 64
automation, 125
border drones, 196
cells on wheels (COWs), 83
Certificates of Waiver or Authorization (COAs), 54
commercial drones, 75, 85
cooperative federalism, 129–30
delivery, 79
domestic military flights, 200–203
drone programs, 115
drones as aircraft, 55–56
expanded operations, 19–20, 20f, 44–47, 182–83
FAA Modernization and Reform Act of 2012 (FMRA), 9–12, 56–57, 212n13
federal agencies, 203–4, 206
federal preemption, 38–41
flights over people, 20–22
Integration Pilot Program, 27–30
line of sight requirements, 85, 94–96
low altitude authorization and notification capability (LAANC), 30–32
media and entertainment, 84
Part 135, 35–37
pilot certification, 14–15, 21, 53, 85

privacy, 228n57
problems with drone policy, 85–86
Reauthorization Act of 2018, 32–35, 218n109
registered drones, 53
remote ID, 23–25, 41–44, 63, 226n28
sandbox program, 102–3, 105
Section 333, 10, 12–14, 56–58
Section 336, 11–12, 25–26, 225n10
Section 2209, 47–48
Section 44807, 61–62, 226n25
spectrum, 50
state and local governments, 66–67
trespass, 70
UAS integration, 5–10
UTM services, 48–49, 65
waiver process, 16–19

Federal Aviation Agency. *See* Civil Aeronautics Authority (CAA)
Federal Communications Commission (FCC), 129
federal government
airspace access, 113
airspace management, 120–25
authority, 113, 115, 128–29
border drones, 194–95
cooperative federalism, 129
domestic drones, 193–94
federal agencies, 203–7
Integration Pilot Program (IPP), 27–30
nonsurface airspace, 117
preemption, 38–41, 66–69, 107–8, 127
privacy, 100–102
remote ID, 23
state and local challenge, 38–41

Federal Highway Administration, 127–28
federal interaction, 108
federalism, 113–14, 126, 145, 159
Federalist Papers, 143, 251n97
federal-market model, 115–16, 243n104
Federal Register, 13, 25–26, 28–29, 31, 37, 43, 212–13n32
federal role in regulation, 113
FedEx, 16
Feinstein, Dianne, 203
Ferguson, Missouri, 204–5
Fifth Amendment, 159
film industry, 84
Financial Conduct Authority, 104
fintech sandbox, 104–6
firefighting, 59, 82
First Amendment, 39, 68, 174, 204
First Institute of the Laws of England (Coke), 132–33
Flight Manual, 45–46
flights over people. *See* operations over people (OOP)
FMRA. *See* FAA Modernization and Reform Act (FMRA)
Fourth Amendment, 1–2, 82
Freedom of Information Act (FOIA), 200–201, 205
FRIAs. *See* FAA-recognized identification areas (FRIAs)

GAO. *See* U.S. Government Accountability Office (GAO)
Gasser, Urs, 125–26, 242n80
General Act Relative to Aeronautics, 153
General Atomics, 196
General Law 534 of 1922, 149
George Washington University, 83
Gibbons v. Ogden, 145–46
Glatt v. Page, 158–59
globe aérostatique, 134
Glossa Ordinaria (Accursius), 131
Gorgon Stare, 184–86, 200
government uses of drones, 99–102. *See also* federal government
Grand Forks Air Force Base, 203
Griggs v. County of Allegheny, 69, 120, 160, 240n52
ground-based delivery vehicles, 80
ground sample distance (GSD), 186
ground surveillance systems, 198
Guarantee Clause, 144
Guardian drones, 194

Hamilton, Alexander, 143
Hawkins, Sir Henry, 136
Hazeltine, Harold D., 237n25, 238n36
Herrin v. Sutherland, 140
Hickenlooper, Smith, 155–56
high-altitude airspace, 115–16
high-rise construction innovation, 118
hobbyists, 11, 25–27, 33–34, 54, 216n83
Holmes, Oliver Wendell, Jr., 140–41
Hoover, Herbert, 121
House Bill 243, Privacy Protection Amendments, 101–2
House of Lords, 136
H.R. 2997, 86
H.R. Rep. No. 69-572, 254n159, 255nn163–64

Huerta, Michael, 22–23, 63
Hurricane Maria, 83

identification (ID), 2, 63
IG. *See* inspector general (IG)
Illinois, 168
industry guidance, 93
infrastructure, 47–48, 80
innovation, 87, 106, 108, 151
inspection laws, 146, 151
inspector general (IG), 206
institutional soundness, 107
insurance companies, 59
integration into national airspace
  aerial corridors, 112–13
  BEYOND program, 103
  detect-and-avoid (DAA), 63–64
  domestic military flights, 201–3
  enabling legislation, 9–12
  expanded operations, 19–20, 20f, 44–47
  FAA Modernization and Reform Act of 2012, 56, 201
  government use of drones, 206–7
  Integration Pilot Program, 28–29, 102–3
  regulatory progress, 5–8
  remote ID, 41–44
  sandboxes, 105
  Section 333, 10, 12–14
  Small UAS ARC, 21
  third-party service supplier model, 32
  UTM, 48–49, 65
Integration of Civil Unmanned Aircraft Systems into the National Airspace System, 9–10
Integration Pilot Program (IPP), 27–30, 37, 40, 102–3
intentional tort, 73
Inter-American Commercial Aviation Convention, 152, 255n168
interconnected networks, 125–26
International Air Navigation Convention of 1919, 148
international air travel, 152
International Association of Chiefs of Police, 193
International Convention Relating to the Regulation of Aerial Navigation, 152, 255n165
Interstate Commerce Commission, 259n254
interstate drone commerce, 129, 259n254
intrusion upon seclusion, 73
IPP. *See* Integration Pilot Program (IPP)

Jome, Hiram, 154
Journal of Air Law and Commerce, 156
judicial authorization, 167–68, 176
jurisdiction, 67, 115, 127, 143–45
justification, 174–75
Justinian law, 131–32

Kavanaugh, Brett, 26
Kent, James, 138, 246–47n52
Kenyon v. Hart, 135
Kestrel system, 197
King, Martin Luther Jr., 172
Kohler, Gramazio, 78

LAANC. *See* low altitude authorization and notification capability (LAANC)
labeling, 92
landing fees, 124
land ownership
  ad coelum, 131–42, 153–55
  airspace as property, 117–20
  avigation easements, 98–99
  drone highway use, 127–28
  property rights, 98, 133–36, 139–47, 153–56, 159–61
  trespass by overflights, 69–72
  Uniform Aeronautics Act, 148–50
law enforcement agencies
  armed drones, 192–93
  border drones, 199
  defense-and-security use of drones, 81–82
  electronic signals collection, 183–84
  emergency exceptions, 175–77
  exhaustion requirements, 170–72
  FAA deference, 204–5
  loan-a-drone programs, 165
  logistical requirements, 174–75
  minimization rules, 172–74
  state and local governments, 66
  sunset provisions, 179
  surveillance, 81–82, 100–101, 165–66, 170–79
The Law of Air (Hazeltine), 237n25, 238n36
The Law of Municipal Corporations (McQuillin), 142
lawsuits, 98, 99, 128, 243n100
L-band spectrum, 50
LEAP. *See* Long Endurance Aircraft Platform (LEAP)
leased airspace, 127–28
Lee, Frederic P., 239n42

Lemmon v. Webb, 136
liability, 149
Libertas Institute, 88, 231n68
licensing and permits, 93
limited remote ID, 41–42
line of sight, 94–96
livestock, 158–59
loan-a-drone programs, 165
Locomotive Act of 1865, 75–76
logistical information, 165–66, 174–75
long-distance operations, 114–15
Long Endurance Aircraft Platform (LEAP), 81, 91
Los Angeles Times (newspaper), 181
low-altitude airspace, 96–97, 112–13, 117–20, 127, 159, 239n42, 240n53
Low Altitude Authorization and Notification Capability (LAANC), 18, 30–32, 48–49, 217n59
Lyman v. Bennet, 139

Madison, James, 143
marine navigation, 151
market allocation, 115
market disposition, 120
Market Empowerment Framework, 88–94, 100
market reforms, 124–25
Marshall, John, 146
Martec's Law, 76, 228n7
Massachusetts, 149
Massachusetts Constitution, 247n54
Matternet, 37
Maule, Sir William Henry, 135
McKinsey analysts, 130
McNary-Watres Act of 1930, 121
McQuillin, Eugene, 142
media and entertainment, 84
media literacy, 92
medical supplies, 60–61, 83
Merchandise Mart, 118
Metropolitan West Side Elevated R.R. Co. v. Springer, 139
Michel, Arthur Holland, 196–97, 199–200
Michigan Supreme Court, 139
micro UAS, 20–22
Micro UAS ARC, 21–22, 44
Mid-Atlantic Aviation Partnership, 37
minimization rules, 165–66, 172–74
model aircraft, 6, 11, 24–27, 34, 54–57, 85, 216n81
Model Airport Zoning Act, 158
Montana Supreme Court, 140
Montgolfier, Joseph-Michel and Jacques-Étienne, 134
Montgomery-Cuninghame, Sir William James, 135
Morales, Jimmy, 177
Morgan, Rob, 105
Morgan Stanley, 80
Motor Carrier Act, 259n254
Mueller, Robert, 203
multispectral cameras, 187, 263n27
municipal zoning ordinances, 147

NAS. *See* National Airspace System (NAS)
NASA, 32, 48, 65, 192
National Academies of Sciences, 95–96
National Air Security Operations Centers, 194, 266n69
national airspace. *See* integration into national airspace
National Airspace System (NAS), 5–8, 31, 56, 63–64, 102–3, 105, 112
National Conference of Commissioners on Uniform State Laws, 147–48, 156, 253n136
National Defense Authorization Act of 2018, 9, 27, 57
National Highway System, 128
National Institute of Municipal Law Officers, 158
National Institute of Standards and Technology, 23
National Press Photographers Association, 39
National Press Photographers Association v. McCraw, 67–68
National Telecommunications and Information Administration (NTIA), 228n57
National Transportation Safety Board, 55
natural disasters, 82–83
navigable aircraft, 137
navigable airspace, 7, 67, 69, 119–20, 127, 151–53, 159–60, 240n45, 240n52
navigable waters, 145–46
Nebraska, 158–59
network remote ID, 41–42, 63
New Hampshire, 147
Newsweek (magazine), 200
Newton, Massachusetts, 38–39, 67
New York, 118

New York Supreme Court, 141
New York Times (newspaper), 147
nighttime operations, 18, 44, 46, 182
NIMBY. *See* not-in-my-backyard (NIMBY)
no-bid contracts, 121–22
no-fly zones, 113, 204
nonsurface airspace, 117
nontargets, 170–72
North Dakota, 193, 203, 204–5
Notice of Proposed Rulemaking (NPRM), 14, 20–23, 41–42, 44–47, 49
Notice to Airmen, 13, 14
not in my backyard (NIMBY), 99
NPRM. *See* Notice of Proposed Rulemaking (NPRM)
NTIA. *See* National Telecommunications and Information Administration (NTIA)
nuisance, 72–73, 134–36, 139–40, 142, 150, 153–55, 158–60

Obama, Barack, 81, 124, 228n57
Office of Information and Regulatory Affairs (OIRA), 22, 31
Office of Legal Services Innovation, 106–7
Ohio, 153–55, 256n177
OIRA. *See* Office of Information and Regulatory Affairs (OIRA)
OOP. *See* operations over people (OOP)
operating conditions and limitations, 12–14, 16–17, 19
operations over people (OOP), 20–22, 44–46, 60, 62, 63, 182, 226n28
overflights, 72, 119, 128, 134, 149, 153, 221n151

pacing problem, 112
package delivery, 36–37, 61–62, 78–79
Palfrey, John, 125–26, 242n80
Paperwork Reduction Act of 1995, 31
Part 107
  adoption of, 14–15
  air carriers, 15–16
  BVLOS operations, 18, 60–61, 65, 67
  commercial drones, 59–61, 85
  delivery, 36–37
  detect-and-avoid (DAA), 64
  expanded operations, 19, 27–28, 34, 44–46
  FAA Modernization and Reform Act of 2012, 56
  flights over people, 20–22
  nighttime operations, 46
  remote ID, 63
  Section 44807, 61
  waiver process for, 16–19
Part 135, 35–37, 61–62, 226n24
Part 298, 37
passenger drones, 114, 243–44n104
Pathfinder Program, 28
Pennsylvania Coal Company, 141
Pennsylvania Supreme Court, 144–45
Penruddock's Case, 133
Pentagon, 200–203
permissionless innovation, 87–88, 91
Persistent Surveillance Systems (PSS), 169, 177–78, 184–89
Persistent Threat Detection System, 197
Petition for Rulemaking, 224n205
Phillips, Jesse J., 139
photography, 55–56, 184–87. *See also* aerial imagery
Pickering v. Rudd, 134, 137
pilot certification, 12, 14–15, 21, 53, 85, 158
pilot training, 7, 46, 158
Pirker, Raphael, 55
Pollock, Sir Charles Edward, 135
Pollock, Frederick, 136–37
Portsmouth Harbor Land & Hotel Company (Holmes), 140–41
postmaster general, 121–22
power and utility companies, 59
precautionary principles, 93–94
precision agriculture, 58
Predators, 194–96, 199, 206
prescriptive easement, 128, 243n100
PricewaterhouseCoopers, 76
pricing of airspace, 123–24
privacy, 73, 82, 101–2, 170–72, 181, 198–99, 206, 228n57
Privacy Protection Act, 102, 232–33n93
private companies, 100
privately owned airports, 128
probable cause warrants, 166–70
prohibition, 94
Project Innovate, 104
Project Maven, 188
property rights, 98, 117–20, 133–36, 139–47, 153–56, 159–61
protests, 168, 204–5
Protocol to the International Convention Relating to the Regulation of Aerial Navigation, 152
prototype evaluation, 31–32

PSS. *See* Persistent Surveillance Systems (PSS)
public acceptance, 66
public highways, 145
public resources regulation, 115–16
public right of way, 142
public use operations, 54
Putney and Wimbledon Commons Act, 136

radio and television networks, 59
real estate law, 147
real-time tracking of drone flights, 182
reckless operation, 11
reconnaissance, 81, 100, 165
reductio ad absurdum, 134
registration requirements, 25–27, 39, 43, 57, 67, 216n83
regulations. *See also* Part 107
  air carriers, 15–16, 46
  Air Commerce Act of 1926, 153
  aircraft certification, 64
  airspace, 117–19, 127–29
  Civil Aeronautics Act, 157–58
  critical infrastructure, 47–48
  drone technology, 76
  enabling legislation, 9–12
  expanded operations, 19
  federal-administrative model, 129–30
  Federal Aviation Act, 67–68
  Market Empowerment Framework, 88–95
  models of, 115–16
  pacing, 112–13
  permissionless innovation, 87–88
  problems with drone policy, 85–86
  public highways, 161–63
  remote ID, 41–44
  sandboxes, 102–8
  Section 333, 10, 12–14, 58
  Section 336, 25–27, 57
  state and local governments, 38–41
  state-market model, 126–29
  UAS integration, 5–8
regulatory sandboxes, 91, 102–8
Remote ID and Tracking ARC, 23–27, 215n73
remote identification (ID), 22–27, 32–34, 41–44, 49, 63
remotely piloted aircraft system (RPAS), 224n1
remote pilot certification process, 14–15
resiliency, 94
Restatement of Property, 40
Restatement of the Law, Fourth, Property, 72
Restatement of the Law, Second, Torts, 68–70, 72
Restatement of the Law, Third, 73
right in land, 133
right of peaceful enjoyment of property, 133–35
Rinaudo, Keller, 111
roadway airspace leasing, 127–28
Roosevelt, Franklin, 121–22
RTCA, 23
Rugg, Arthur, 149–50

sandbox programs, 91, 102–8
Schuler, Ari, 196
search-and-rescue operations, 29, 58, 84, 191
Seattle, Washington, 199–200
secretary of transportation, 10–11
Section 91.137, 61, 225–26n21
Section 107.1, 15–16
Section 107.29, 18
Section 107.41, 18
Section 158, 68
Section 159, 69–70
Section 332, 9, 12, 14, 212n13
Section 333, 10, 12–13, 14, 17–18, 36, 56–58, 60, 61
Section 336, 11–12, 24–27, 33–34, 56–57, 225n10
Section 349, 34, 218n108
Section 376, 48–49
Section 2202, 23
Section 2208, 48
Section 2209, 47–48
Section 44807, 36, 61–62, 63, 65, 226n25
Section 652A, 73
Section 652B, 73
Section 821D, 72
see and avoid. *See* detect-and-avoid (DAA)
self-defense, 152
self-regulation, 91–92
Sierra Nevada Corporation, 200
Silverthorne, Colorado, 235n12
Singapore, 104
Singer v. Newton, 67
Single Pilot certificates, 36
SkyGuardian, 196
Skyward, 31–32

slot auctions, 124
slot rationing, 123–24
Small UAS ARC, 21–22
Small UAV Coalition, 60–61, 225n21
small unmanned aerial systems (sUAS), 6–7, 56, 65, 195–96
Smith v. New England Aircraft Co., 149–50
social norms and pressures, 91
solar-powered drones, 192
SOSC. *See* Systems Operations Support Center (SOSC)
Sotomayor, Sonia, 167
South Korea, 104
sovereignty, 127, 143–45, 148–52, 157, 160–63, 239–40n42
Special Committee F-38, 63–64
Special Government Interest authorizations, 204
spectrum, 49–51
stakeholder concerns, 22–25, 26, 33, 40, 45
standard remote ID, 41–44
Stanley, Jay, 169
state and local governments
  Air Commerce Act of 1926, 151–56
  air rights, 98
  airspace access, 112–13
  airspace management, 120
  authority, 115, 128–29, 161
  avigation easements, 99
  Civil Aeronautics Act, 157–58
  FAA guidance, 38–41
  federal preemption, 66–68
  Integration Pilot Program (IPP), 30
  privacy, 73
  property rights, 143–47, 156
  sandboxes, 102–8
  Section 2209, 47–48
  sovereignty, 143–45, 148–52, 157, 160–63
  state preemption, 97–98
  surface airspace claims, 118–20
  surveillance, 166
state-market model, 115–16, 126–29
Stingrays. *See* cell-site simulators
sUAS. *See* small unmanned aerial systems (sUAS)
sunset provisions, 48, 65, 178–79, 261n25
Supreme Court
  ad coelum, 69, 141–42, 159, 236n22
  armed drones, 192–93
  model aircraft, 57
  navigable airspace, 69, 119–20, 159–60
  state sovereignty, 144–47, 161–63
  surface airspace, 119–20
  surveillance, 1–2, 167–68
  Takings Clause, 240n51
  trespass, 69, 140–41
Supreme Court of Iowa, 142
Supreme Court of Massachusetts, 149–50
surface airspace, 96–97, 112–13, 117–20, 127, 239n42
surveillance
  autonomous flight, 191–92
  border drones, 197–98
  camera technology, 184–87
  defense and security, 81–82
  development of drone operations, 54–55
  domestic military flights, 199–200
  drones as platform for, 183–84
  electronic means vs. human activity, 167–68, 260n7
  emergency exceptions to, 175–77
  exhaustion requirements, 170–72
  expanded operations, 181–83
  federal agencies, 205–6
  Fourth Amendment, 1–2
  government use of drones, 81, 100, 165–66
  law enforcement agencies, 81–82, 100–101, 165–66, 170–79
  legislation, 177–79
  limits to, 165–66
  logistical requirements for, 174–75
  minimization rules, 172–74
  no-fly provision, 67–68
  probable cause warrants, 166–70
  sunsets, 178–79
  video analytics, 187–90
  warrantless drones, 1–2
  wide-area aerial surveillance (WAAS), 184–88, 196–98, 206
  Wide Area Surveillance System (WASS), 197
survey and inspection operations, 58
Swetland, R. H., 153
Swetland v. Curtiss Airports Corporation, 153–55
synthetic aperture radar, 194–95, 200
Systems Operations Support Center (SOSC), 204

Takings Clause, 240n51
Taylor, John, 26, 57, 216n81
Taylor v. Huerta, 26, 57, 216n81
technology advancement, 111–12, 182–83

technology imperatives. *See* aircraft certification; detect-and-avoid (DAA); remote identification (ID); UAS Traffic Management (UTM)
technology lock-in, 125–26
technology use cases, 76–84
telecommunications, 129
Telegraph Act of 1863, 236n22
Tennessee v. Garner, 192–93
Tenth Amendment, 144
terminal access, 123–25
tethered aerostat, 177–78
Tethered Aerostat Radar System program, 197
Texas Government Code, 67–68
The Guardian, 200
Thierer, Adam, 88
third-party service supplier model, 32
tort law, 66, 68–72, 136
tracking technology, 23–25, 167–68, 171
Transcontinental & Western Air (TWA), 122
transparency, 92, 100, 107
Transportation Security Administration, 14–15
trespass, 39–40, 68–72, 99, 133–37, 139–42, 149–50, 153–56, 158–60
Trump, Donald, 200–201
TWA. *See* Transcontinental & Western Air (TWA)
12th Middlesex Volunteer Corp, 135–36

UAS. *See* unmanned aircraft system (UAS); drones
UAS Identification and Tracking Aviation Rulemaking Committee, 63
UAS integration, 33f
UAS service supplier (USS), 31–32, 41–42
UAS Traffic Management (UTM), 32, 42, 44, 48–49, 65, 128, 130, 244n106
UAV. *See* unmanned aerial vehicle (UAV)
ULC. *See* Uniform Law Commission (ULC)
ULC Plenary, 70–72
Ultra LEAP, 81, 191
Uniform Aeronautical Regulatory Act, 156
Uniform Aeronautics Act, 148–50, 253n137, 257n212
Uniform Law Commission (ULC), 39–40, 70, 72
Uniform Licensing Act, 156
Uniform State Law for Aeronautics, 238n37
Uniform Tort Law Relating to Drones Act, 70
unitary federal authority, 113
United Aircraft & Transport, 122
United Kingdom, 75–76, 104, 236n22
United Parcel Service (UPS), 37, 112
United States v. Causby, 69, 114, 119–20, 159–60, 221n151, 236n22, 240n49, 240nn45–46
United States v. Jones, 167
University of Virginia, 55, 132
unmanned aerial vehicle (UAV), 5–7, 22
unmanned aircraft (UA), 6–7, 222n7
unmanned aircraft system (UAS), 22, 183
unmanned traffic management (UTM). *See* UAS Traffic Management (UTM)
unpiloted remote-control aircraft. *See* model aircraft
UPP. *See* UTM Pilot Program (UPP)
UPS. *See* United Parcel Service (UPS)
UPS Flight Forward, 36–37, 61–62
urban air mobility (UAM), 47, 129–30
U.S. Air Force, 191, 201
U.S. Coast and Geodetic Survey, 254–55n161
U.S. Court of Appeals for the D.C. Circuit, 26–27, 188
U.S. Court of Appeals for the Sixth Circuit, 155
U.S. criminal code, 35
U.S. Customs and Border Protection (CBP), 181–82, 193, 194–99, 206
U.S. Department of Justice (DOJ), 35, 206
U.S. Department of Transportation (DOT), 23, 27–29, 37, 40, 46, 70, 102–3, 202
U.S. Department of Transportation Office of Inspector General, 18–19
U.S. District Court for the District of Massachusetts, 39
U.S. District Court for the Western District of Texas, 39
U.S. Government Accountability Office (GAO), 124, 195, 201, 227n37
U.S. military, 54–55, 69, 81–82, 100, 119, 165, 184–91, 199–203
U.S. Ninth Circuit Court of Appeals, 16
USA FREEDOM Act, 261n25
USA PATRIOT Act, 261n25
USA Today (newspaper), 200

USS. *See* UAS service supplier (USS)
Utah, 101, 104–7, 191
UTM. *See* UAS Traffic Management (UTM)
UTM Pilot Program (UPP), 48–49
UTM system pilot program, 65

VADER. *See* Vehicle and Dismount Exploitation Radar (VADER)
Vehicle and Dismount Exploitation Radar (VADER), 194–97
Vermont, 138
video analytic tools, 169–70, 187–90, 196
Virgil, 131
Virginia, 112, 138
visual line of site (VLOS), 20f

waiver process, 17–19, 59–61
Waiver Safety Explanation Guidelines, 17
WakeMed hospital, 37
Wall Street Journal (newspaper), 124
WAMI. *See* wide-area motion imagery (WAMI)
war power clause, 148
warrantless drone surveillance, 1–2
Washington Post (newspaper), 201
WASS. *See* Wide Area Surveillance System (WASS)
weapons, 81, 100, 193
Wheeler v. City of Ft. Dodge, 250–51n89
White House memorandum, 43–44
wide-area aerial surveillance (WAAS), 184–88, 196–98, 206
wide-area motion imagery (WAMI), 184–87, 189–90, 196–97, 200
Wide Area Surveillance System (WASS), 197
Williams, Jim, 132
Wimbledon Common, 135–36
Wing Aviation, 36–37, 61–62, 79–80
Winston, Clifford, 125
Wiretap Act, 170–74
wiretaps, 170, 172, 174, 260n14
The Wire (television show), 260n14
Wright brothers, 256n177
Wyoming, 153

Zenko, Micah, 81
Zipline, 111

ABOUT THE EDITOR

MATTHEW FEENEY is the director of the Cato Institute's Project on Emerging Technologies, where he works on issues concerning the intersection of new technologies and civil liberties. Before coming to Cato, he worked at *Reason* magazine as assistant editor of Reason.com. He has also worked at *The American Conservative*, the Liberal Democrats, and the Institute of Economic Affairs. His writing has appeared in the *New York Times*, the *Washington Post*, *HuffPost*, *The Hill*, the *San Francisco Chronicle*, the *Washington Examiner*, *City A.M.*, and others. Feeney also contributed a chapter to Libertarianism.org's *Visions of Liberty*. He received both his BA and MA in philosophy from the University of Reading.

ABOUT THE CONTRIBUTORS

SARA BAXENBERG is an associate at Wiley Rein, where she offers counsel to telecommunications and emerging technology companies on a wide variety of regulatory, litigation, and compliance matters. Her areas of focus include unmanned aircraft systems, telecommunications infrastructure deployment, communications-related disability access law, social media and content moderation, and wireless and satellite regulatory issues. She has written on drone law and policy for publications including *Law360*, the *Westlaw Journal Aviation,* the American Bar Association's *Air & Space Lawyer,* and the Denver Bar magazine. Baxenberg earned her BA from Brown University and her JD from Harvard Law School.

JAMES CZERNIAWSKI is a policy analyst at Libertas Institute, a free-market think tank in Utah. He writes about consumer data privacy, cybersecurity, and technology and innovation issues. He earned his MA in economics from George Mason University, where he was an MA fellow at the Mercatus Center. He graduated from the State University of New York-Purchase with a BA in economics and American history. His work has been published in *Real Clear Future*, the *Morning Consult*, *Deseret News*, the *National Interest*, the *Salt Lake Tribune*, the *Washington Examiner*, and many others.

LAURA K. DONOHUE is a professor of law at Georgetown Law, director of Georgetown's Center on National Security and the Law, and director of the Center on Privacy and Technology. Her work on new and emerging technologies centers on social media, biometric identification, augmented and virtual reality, artificial intelligence, and drones. She is the author of *The Future of Foreign Intelligence: Privacy and Surveillance in a Digital Age* (Oxford University Press, 2016), *The Cost of Counterterrorism: Power, Politics, and Liberty* (Cambridge University Press, 2008), and *Counterterrorist Law and Emergency Law in the United Kingdom, 1922–2000* (Irish Academic Press, 2007). Donohue obtained her JD from Stanford Law School and her PhD in history from the University of Cambridge.

JAKE LAPERRUQUE is senior counsel at The Constitution Project at the Project on Government Oversight (POGO). He oversees work on privacy, surveillance, and cybersecurity issues, highlighting how emerging technologies are rapidly impacting Constitutional rights and principles. His work focuses on foreign intelligence surveillance, location privacy, cellphone privacy, facial recognition, aerial surveillance, and election security. Previously, Laperruque worked as a fellow on privacy, surveillance, and security at the Center for Democracy & Technology and as a program fellow at the Open Technology Institute. He also served as a law clerk on the Senate Subcommittee on Privacy, Technology, and the Law. He is a graduate of Harvard Law School and Washington University in St. Louis.

BRENT SKORUP is a senior research fellow at the Mercatus Center at George Mason University. His research areas include transportation technology, telecommunications, aviation, and wireless policy. He serves on the Federal Communications Commission's Broadband Deployment Advisory Committee and on the Texas Department of Transportation's Connected and Autonomous Vehicle Task Force. He is also a member of the Federalist Society's Regulatory Transparency Project. Before joining Mercatus, he was the director of research at the Information Economy Project, a law and economics university research center. Skorup has a BA in economics from Wheaton College and a law degree from the George Mason University School of Law.

JAY STANLEY is a senior policy analyst with the American Civil Liberties Union (ACLU) Speech, Privacy, and Technology Project, where he researches, writes, and speaks about technology-related privacy and civil liberties issues and their future. He is the editor of the ACLU's Free Future blog and has authored and coauthored a variety of influential ACLU reports on privacy and technology topics. Before joining the ACLU, he was an analyst at the technology research firm Forrester, served as American politics editor of Facts on File's World News Digest, and as national newswire editor at Medialink. He is a graduate of Williams College and holds an MA in American history from the University of Virginia.

GREGORY S. WALDEN is a partner at Dentons US LLP. Walden serves as aviation counsel for the Small UAVE Coalition and is an adjunct professor at the Antonin Scalia Law School at George Mason University. He teaches courses in aviation law and automated vehicles law. Walden previously served as chief counsel of the Federal Aviation Administration and as associate deputy attorney general with the U.S. Department of Justice, where he was also special assistant and counselor to the assistant attorney general of the Civil Division. He holds a JD from the University of San Diego School of Law and a BA from Washington & Lee University.

ABOUT THE CATO INSTITUTE

Founded in 1977, the Cato Institute is a public policy research foundation dedicated to broadening the parameters of policy debate to allow consideration of more options that are consistent with the principles of limited government, individual liberty, and peace. To that end, the Institute strives to achieve greater involvement of the intelligent, concerned lay public in questions of policy and the proper role of government.

The Institute is named for *Cato's Letters*, libertarian pamphlets that were widely read in the American Colonies in the early 18th century and played a major role in laying the philosophical foundation for the American Revolution.

Despite the achievement of the nation's Founders, today virtually no aspect of life is free from government encroachment. A pervasive intolerance for individual rights is shown by government's arbitrary intrusions into private economic transactions and its disregard for civil liberties. And while freedom around the globe has notably increased in the past several decades, many countries have moved in the opposite direction, and most governments still do not respect or safeguard the wide range of civil and economic liberties.

To address those issues, the Cato Institute undertakes an extensive publications program on the complete spectrum of policy issues. Books, monographs, and shorter studies are commissioned to examine the federal budget, Social Security, regulation, military spending, international trade, and myriad other issues. Major policy conferences are held throughout the year, from which papers are published thrice yearly in the *Cato Journal*. The Institute also publishes the quarterly magazine *Regulation*.

In order to maintain its independence, the Cato Institute accepts no government funding. Contributions are received from foundations, corporations, and individuals, and other revenue is generated from the sale of publications. The Institute is a nonprofit, tax-exempt, educational foundation under Section 501(c)3 of the Internal Revenue Code.

Cato Institute
1000 Massachusetts Avenue NW
Washington, DC 20001
www.cato.org